Events Tourism

This book presents critical insights and contemporary perspectives for exploring current trends, concerns and prospects of events tourism. It examines modern-day global issues facing the events and tourism industry, policymakers, researchers and academics to advance understanding of practice and development of theory.

Organised in four parts, this book examines how events tourism is designed, planned and delivered. The first part engages with the core, fundamental concepts of events tourism which establish a basic understanding of the field. The second part addresses contemporary issues related to visitor attractions, music festivals, small and user-generated events, wanderlust and entrepreneurship. The third part focuses on meetings and challenges in the conference industry after disasters, the economic impact and other dilemmas of mega-events, and city and destination concerns. The fourth and final part provides a peek into the future of events tourism vis-à-vis reshaping cities, music festivals and critical dilemmas of the 21st century.

With an international appeal because of cross-national contributions, this book will interest events and tourism practitioners, academics, students, researchers, policymakers, and business and investment sector professionals across the globe.

Violet V. Cuffy is a senior lecturer in events and tourism management at the University of Bedfordshire with a background in tourism consultancy. Violet is co-editor of *Lifelong Learning for Tourism: An International Perspective* and principal investigator of Arts Humanities Research Council grant on Dominica Creole heritage.

Fiona Bakas is a critical tourism researcher and part-time lecturer, with events industry experience in the conference sector. Fiona is currently working at Coimbra University, Portugal, as postdoctoral researcher on a nationwide project on creative tourism in rural areas and small cities, called CREATOUR.

Willem J. L. Coetzee is a senior lecturer in tourism at the University of Otago, New Zealand. His research theme is within the nexus of tourism, events and the SDGs, with a focus on mega and hallmark events, and tourism within water-stressed destinations.

Events Tourism

Critical Insights and Contemporary
Perspectives

**Edited by
Violet V. Cuffy, Fiona Bakas
and Willem J. L. Coetzee**

Routledge
Taylor & Francis Group

LONDON AND NEW YORK

First published 2021
by Routledge
2 Park Square, Milton Park, Abingdon, Oxon OX14 4RN

and by Routledge
605 Third Avenue, New York, NY 10017

First issued in paperback 2022

Routledge is an imprint of the Taylor & Francis Group, an informa business

British Library Cataloguing-in-Publication Data
A catalogue record for this book is available from the British Library

Library of Congress Cataloging-in-Publication Data
A catalog record has been requested for this book

ISBN 13: 978-0-367-61642-7 (pbk)
ISBN 13: 978-0-367-36167-9 (hbk)
ISBN 13: 978-0-429-34426-8 (ebk)

DOI: 10.4324/9780429344268

Typeset in Times New Roman
by codeMantra

We dedicate this volume to our chapter contributors who stuck with us to the end to see this volume to completion. A long, but fruitful journey in the end.
Thanks for your dedication and patience.

Contents

List of figures   x
List of tables   xi
List of boxes   xii
Notes on contributors   xiii
Preface   xvii

PART I
Fundamental concepts   1

 1 **Events tourism: an overview**   3
    FIONA BAKAS, WILLEM J. L. COETZEE AND
    VIOLET V. CUFFY

PART II
Contemporary perspectives   11

 2 **Events as Visitor Attractions in Destination
    Shopping Centres**   13
    DAVID STRAFFORD

 3 **Managing the event workforce: analysing
    the heterogeneity of job experiences**   33
    BIRGIT MUSKAT AND JUDITH MAIR

 4 **User-generated events in tourism:
    a new must-not-miss phenomenon**   47
    ESTELA MARINE-ROIG, EVA MARTIN-FUENTES AND
    NATALIA DARIES-RAMÓN

5 **Undocumented! Small events of rural hinterlands, South Africa**   70

UNATHI SONWABILE HENAMA
AND PETRUS MFANAMPELA MAPHANGA

6 **Entrepreneurship development and Slow Food events**   86

BURCIN KALABAY HATIPOGLU, ONNO ANIL,
SAADET MEMIŞ AND DILAN ŞAHIN

7 **Riding wanderlust: the case of motorcycle events**   103

LUCIA CICERO

PART III
Regeneration, displacement and planning frameworks   123

8 **Quake aftermath and conference industry transformation, Christchurch**   125

ABRAR FAISAL, JULIA N. ALBRECHT AND
WILLEM J. L. COETZEE

9 **A critical view on mega-events economic impacts studies: Milan 2015 and beyond**   143

JÉRÔME MASSIANI

10 **Destination Israel and Malawi wanderlusts**   163

JAMES MALITONI CHILEMBWE

11 **City rebranding, social discontent and bidding for cultural events**   184

DANIEL BARRERA-FERNÁNDEZ, MIGUEL ANXO
RODRÍGUEZ-GONZÁLEZ AND MARCO
HERNÁNDEZ-ESCAMPA

12 **MICE tourism development in Ethiopia**   196

GEBEYAW AMBELU DEGAREGE AND BRENT LOVELOCK

PART IV
The future of events tourism   215

13 **Reshaping metropolitan cities and creative tourism through artists' vision**   217

SILVIA GRANDI AND CHIARA BERNASCONI

**14 Counterculture and the future of music festivals
and events** 232

DANIEL WRIGHT

15 Events tourism: a critical debate for the 21st century 251

VIOLET V. CUFFY AND BIPITHALAL
BALAKRISHNAN NAIR

Index 257

Figures

4.1	Unified modelling language (UML) diagram to store and process an event	52
4.2	Followers per user	56
4.3	Posts per user	57
4.4	User profile	57
4.5	Range of likes per post (%)	59
4.6	Range of comments per post (%)	59
6.1	A framework for supporting rural entrepreneurship through a planned event	93
9.1	Various impacts considered in Milan 2015 impact studies (bln euro)	146
9.2	Activation of local expenditure: the spectrum of possible impact magnitudes	150
9.3	Comparison among estimation methods (added value in bln euro)	153
9.4	Value added redistributed by Expo 2015 through changes in expenditures from Italians and foreigners	158
12.1	International tourist arrivals, 2000–2016	199
12.2	International tourism by the purpose of visit, 2010–2015	199
12.3	Growth and contribution of the business purpose of visitors, 2010–2015	200
12.4	International tourist arrivals in Addis Ababa, 2009/2010–2014/2015	201
12.5	Number of star-rated hotels, beds and rooms in Addis Ababa, 2017	205
14.1	The cycle of commodified culture: from counterculture to mainstream culture	240

Tables

3.1	Heterogeneity in work teams	38
4.1	Meetups in the Eastern Pyrenees	54
4.2	Information about Instagram users who posted photographs	56
4.3	Repeat participants to the events	58
4.4	Information by photographs posted related to the events	58
5.1	Key elements for Community-Based Tourism (CBT) success	80
7.1	Motives for attending motorcycle events	107
7.2	Wanderlust scale	109
7.3	Scales on travel motivation	111
7.4	Sample descriptive statistics	112
7.5	Frequency of travelling to motorcycle events and length of stay	113
7.6	Wanderlust scale for attending motorcycle events	114
7.7	Respondents according to wanderlust segments and results of significant tests	115
7.8	Main motives for attending motorcycle events: mean scores (ranks)	116
8.1	Christchurch venues and conference facilities in hotels	130
9.1	Quantification of additional visitors for Italy	148
9.2	An increase in consumption: an overview of possible assumptions	151
9.3	Increased added value in Italy generated by visitors' expenditures (mln €)	152
9.4	Additionality of visitors in BOP study	154
9.5	Additionality of visitors for Lombardy	155
9.6	Increased added value generated by visitors' expenditures in Lombardy (mln €)	157
9A.1	Daily expenditure assumption	160
12.1	MICE destinations in Africa, 2016	202
12.2	MICE cities in Africa, 2016	202
15.1	Key factors affecting future events tourism	252
15.2	Consumer trends shaping events tourism (in the area of events and festival)	255

Boxes

2.1 Industry voice: Alex Caley, marketing and experience
    manager, Meadowhall Sheffield   17
2.2 Industry voice: Nikki Tansey, event manager at Intu
    Trafford Centre   21
2.3 Industry voice: Kim Priest, Marketing Manager at Centre:
    MK in Milton Keynes   27
9.1 The expenditure of foreigners   149
9.2 Computation of the additional expenditure for Lombardy   155

Contributors

Julia N. Albrecht is a senior lecturer in the Department of Tourism, University of Otago. Her research interests include tourism management and strategy, visitor management, guiding and interpretation, and nature-based tourism.

Onno Anıl has graduated from Bogazici University, Tourism Administration program. He works at a technology company in Istanbul, Turkey. Recently, he is attending the postgraduate program of Tourism Administration at Istanbul University.

Daniel Barrera-Fernández is professor in the Faculty of Architecture of the University of Oaxaca (Mexico). He is a delegate of ATLAS (Association for Tourism and Leisure Education and Research) for Mexico, Central America and the Caribbean. His research interests focus on urban planning and cultural tourism.

Chiara Bernasconi is an assistant director in Digital Media at the Museum of Modern Art in New York City. He is also a film producer, curator and lecturer in New Media in the Arts and Cultural Policy at the Milan Catholic University and at the School of the Art Institute of Chicago.

James Malitoni Chilembwe is a lecturer at the Faculty of Tourism, Hospitality, and Management of Mzuzu University, Malawi. He holds a Master of Science in International Tourism Enterprise from Glasgow Caledonian University (GCU). He is a PhD candidate at GCU in the UK.

Lucia Cicero is a lecturer at the Ca' Foscari University of Venice, Italy. She graduated with a thesis on marketing to families in tourism. Her research interests focus on family tourism, food and wine tourism, leisure and sustainable practices both in recreational and business activities.

Natalia Daries-Ramón holds a PhD in Business Administration. Currently she is Coordinator of the Degree in Tourism at the University of Lleida, Spain and assistant professor in Hospitality and Gastronomy at the Open University of Catalonia. Her research focuses on the impact of new technologies in tourism.

Gebeyaw Ambelu Degarege is a lecturer in the Department of Tourism, Sport and Society, Lincoln University, Christchurch, New Zealand. Gebeyaw holds a PhD in Tourism from the University of Otago in New Zealand. Gebeyaw's current research interests lie in the wider field of tourism and sustainable development.

Abrar Faisal is a lecturer in Tourism and Event Management at the Auckland University of Technology. Abrar's current research focuses on urban tourism, placemaking, event impacts and legacy, organisational ecology, entrepreneurial resilience and crisis management.

Silvia Grandi is economic, political geographer and adjunct professor in Cultural Tourism and Recreation Research Management at the University of Salento, and in Geography at the University of Bologna. Publications include tourism and recreation activities. She launched the International Summer School in Geography of Tourism at the Rimini Campus.

Burcin Kalabay Hatipoglu is an assistant professor at Boğaziçi University in Istanbul, Turkey, and a visiting fellow at UNSW Sydney, Australia. She has been involved in several sustainable tourism development projects. Her research interests include sustainability reporting, cross-sector partnerships, women entrepreneurship and sustainable tourism management.

Unathi Sonwabile Henama is a lecturer in tourism marketing at the Department of Tourism Management of the Tshwane University of Technology. He has a master's degree specializing on Africa Studies. He has written numerous articles and presented research papers at both local and international conferences.

Marco Hernández-Escampa is a professor in the Faculty of Architecture of the University of Oaxaca (Mexico). He is a delegate of ATLAS (Association for Tourism and Leisure Education and Research) for Mexico, Central America and the Caribbean. His research interests focus on urban anthropology and heritage conservation.

Brent Lovelock is based at University of Otago, Department of Tourism in Dunedin, New Zealand. Brent's research focuses on the application of sustainable tourism approaches: environmentally, socially and politically. He has undertaken research in North America, New Zealand and the Asia-Pacific region, examining collaborative planning processes for sustainable tourism development.

Judith Mair is an associate professor and discipline leader of the Tourism Discipline Group in the UQ Business School, Brisbane, Australia. Her research interests include the impacts of events on community and society, and consumer behaviour in events and tourism.

Petrus Mfanampela Maphanga is a technical assistant in the Department of Tourism Management at the Tshwane University of Technology. He has a Masters in Human Resource Management specializing on HR Studies. He has written numerous research papers and articles.

Estela Marine-Roig holds a European PhD in Tourism. Estela is currently a Serra Húnter Fellow and Coordinator of Social Media Marketing (MSc) at the University of Lleida, Catalonia; Assistant Professor of Social Media and Smart Tourism (MSc) at the Open University of Catalonia; and IFITT Board member serving as Director of Research Excellence.

Eva Martin-Fuentes holds a PhD in Engineering and Information Technologies (Excellent Cum Laude) awarded by the International Federation for Information Technologies and Travel and Tourism IFITT. She is a professor at the University of Lleida, recognized with the Teaching Excellence Award. Currently, she is vice-dean of Institutional Relations and Employability.

Jérôme Massiani is researcher at Ca' Foscari University. He worked for 10 years in a Dutch and French consultancy for economic forecast and programming. His PhD is in economic evaluation applied to transports. He has participated in various research projects in Germany and Italy collaborating with the Universities of Bergamo, Trieste and Venice.

Saadet Memiş has graduated from Bogazici University, Tourism Administration program with a high honour degree. She works for a global technology company as a sales and administrative support in Istanbul.

Birgit Muskat is a senior lecturer in the Research School of Management at the Australian National University. Birgit's research interests include tourist experience management, entrepreneurship and innovation. Her research focuses on the services sector, specifically tourism, hospitality and events.

Bipithalal Balakrishnan Nair is a transdisciplinary researcher in Tourism Studies. She has doctorate in tourism management from the University of Bedfordshire, UK. Dr Nair currently works as an assistant professor in the Department of Tourism and Hospitality, Woosong University, South Korea.

Miguel Anxo Rodríguez-González is a professor in the Department of Art History, University of Santiago de Compostela. He is specialized in Art Theory and Contemporary Art. His research interests focus on current art since mid-20th century, sculpture, photography, contexts of art production and cultural policies.

Dilan Şahin has graduated from Bogazici University, Tourism Administration program. She lives and works in the United States of America. She has internship experiences at five-star hotels in Istanbul and also in the finance industry.

David Strafford is a senior lecturer in Events Management at Sheffield Business School, Sheffield Hallam University, and he is a director of the experiential events agency Hopper Events Ltd. He is working towards a PhD, examining the strategic use of events by Destination Shopping Centres.

Daniel Wright is a lecturer at the University of Central Lancashire. He is an active researcher, educator, publisher and presents at international conferences. His research explores the evolving nature and potential future of the tourism industry.

Preface

The discourse on events tourism as a distinct area is advancing and gaining much more attention in current literature, especially because the sector represents an industry exhibiting exponential growth since 2008 (Getz & Page, 2016b). Events tourism is an increasingly powerful marketing tool and widely employed to enhance destination attractiveness. Up to now, prevalent research focuses on theory and policy for planned events (Getz & Page, 2016a), religious tourism and pilgrimage management (Raj & Griffin, 2015), large-scale sporting events and "mega-events" (Kim, Jun, Walker, & Drane, 2015). However, limited literature has approached the subject of events tourism from the critical perspective of regeneration and development. Meanwhile, the body of critical debate is advancing particularly in tourism research with an increasing number of the academic community recognising the essentiality of critiquing existing theories, conceptual frameworks and ways of knowing (Costa et al., 2017; Ren, Pritchard, & Morgan, 2010; Tribe, 2008). To date, exploration of critical issues specifically to events tourism, regeneration and development in the 21st century remains under-researched, with sparse empirical evidence and limited use of supporting relevant conceptual frameworks.

Further, current gaps include a rich body of work focused on issues such as the socio-cultural and environmental impacts of events and the relationship between events governance and public policy agendas (Mair & Whitford, 2013), among other developing trends. The main objective of this book is to address the knowledge gap in this area, the fundamental and diverse aspects of the events tourism sector, and key academic debates. In that regard, the volume further explores these concepts under the following four main themes:

- Part I: Fundamental concepts
- Part II: Contemporary perspectives
- Part III: Regeneration, displacement and planning frameworks
- Part IV: The future of events tourism

Events Tourism: Critical Insights and Contemporary Perspectives presents a wide range of theoretical and contemporary perspectives underpinning events tourism. This book focuses on global issues faced by the events industry, policymakers, researchers and academics in advancing the understanding of practice and development of theory.

The book is organised in distinct parts around four main themes. The first part establishes the core and fundamental concepts of events tourism with a view to establish basic yet pertinent understanding of the field. The second part focuses on contemporary perspectives from a destination and stakeholder standpoint through the lenses of key sub-niches. The third part presents critical insights into issues related to the conference industry after disasters; regeneration and displacement issues; events tourism as a tool of cultural preservation; impacts, bidding and city branding; efficiency issues in the MICE market; and entrepreneurship. The final part of this book provides a gaze into the future of events tourism, signposting key critical dilemmas of the 21st century.

The value of this book is in acknowledgement of the significance of the sector and its continued growth over time as a key contributor to national development and global social, cultural and economic advancement. Importantly, events tourism has gained prominence as a strategic tool for sustainable development and furtherance of Intangible Cultural Heritage precepts worldwide.

Violet V. Cuffy
Fiona Bakas
Willem J. L. Coetzee

Part I

Fundamental concepts

1 Events tourism

An overview

Fiona Bakas, Willem J. L. Coetzee
and Violet V. Cuffy

Introduction

"The term events tourism was not widely used prior to 1987" (Getz and Page, 2016b: 597). This introductory chapter offers an overview of the core theoretical framework underpinning events tourism, approaching it both as a professional practice and a field of academic study. Moreover, it was only a few decades ago that "events tourism" became established in the tourism industry and research community, with the classification of three main types: business events (e.g. conferences), festivals (e.g. music festivals) and sports (e.g. Olympics). The core proposition defined is one of an instrumental subfield (Getz and Page, 2016b: 597).

Now, "events tourism" is recognised as being inclusive of all planned events and of having an integrated approach to development and marketing (Getz, 2008). Getz argues that past research focused on defining the term "events tourism" and the economic impact of events in the 1970s; the supply and demand side of events and the destination perspective of events in the 1980s; management of events in the 1990s; and the social and cultural impacts of events in 2000. Currently, much research is separated by event type. The uniqueness of this type of tourism is centred around events being spatial-temporal phenomena which are never the same, with live attendance in person essential in order to experience what is marketed as a "one-off" experience. An event then, according to Zizek (2014), is an "amphibious notion with more than 50 shades of grey". It is something that happens beyond sufficient reason and has the potential to change the perception of reality. Furthermore, the event phenomenon is inherent to humans and society, and thus plays a significant part in structuring and maintaining societies (Pernecky, 2013).

Events tourism has progressively gained attention from tourism academics, especially because it is an industry exhibiting exponential growth since 2008 (Getz and Page, 2016b). They are increasingly becoming more powerful marketing tools and are now central to advancing a destination's attractiveness. Gaps in events research have been identified in various areas such as the socio-cultural and environmental impacts of events and the relationship between events and public policy agendas (Mair and Whitford, 2013). To some extent these gaps have been addressed by recent literature that focuses predominantly on theory and policy for planned events (Getz and Page, 2016a), religious tourism and pilgrimage management (Raj and Griffin, 2015) and large-scale sporting events or "mega-events" (Kim et al., 2015). Much events tourism theorisation regarding tourist motivation draws on social psychology to explain why tourists choose to visit the events they do and what that means to them. As much events tourism research has taken an economic impact angle; nevertheless, economic theorisation also plays an important role within events tourism. Nowadays there is a concerted need to examine outcomes and impacts at the personal, political and societal levels, and also in terms of cultural and environmental change (Getz and Page, 2016a). Due to the complexity of events tourism and its effects on society, the journey to establish the theory and practices associated with contextualising events within critical event studies is compared to an "odyssey" (Spracklen and Lamond, 2016). This book aims to address the above gaps in events tourism literature and theory, presenting new empirical research from a wide geographic scope, critically analysing contemporary issues within events tourism.

Adopting a critical approach is now an imperative for tourism research, with an increasing number of the academic community recognising the essentiality of critiquing existing theories, conceptual frameworks and ways of knowing (Tribe, 2008; Ren, Pritchard and Morgan, 2010; Costa et al., 2017). When planning for means by which events can leverage greater community benefit and hence potentially increase social-cultural sustainability, it should not be assumed that events are uncontested and accepted as positive opportunities (Chalip, 2006). For example, whilst festivals are a source of community pride and an expression of shared values and identity, the accrual of social capital may be uneven thus exacerbating existing inequalities, as shown in a research on two festivals in East London (Stevenson, 2016). Moreover, there is a growing body of literature exploring the potential of events to develop social capital (Pernecky, 2013). Further, social capital can be defined as the features of social life (networks, norms and trust) that enable participants to pursue shared objectives and is the force that binds society together, transforming egocentric individuals into members of a community with shared interests and is linked to co-operation and reciprocity (Ziakas, 2016).

Current critical event studies are searching for a richer understanding of what is understood as an event and exploring how they are historically, culturally and politically produced (Hannam, Mostafanezhad and Rickly, 2016). Moving away from a focus on operationalist concerns, critical event studies are taking cultural, political and social trajectories (Lamond and Platt, 2016). These studies identify an urgent need to go beyond the assumption that event evaluations are fundamentally economic impact assessments and to practice critical reflexivity in the study of events. Significant current research on critical events tourism points to the importance of questioning whether local communities are consulted in event planning that takes place in their city or town. Whilst previously, the enjoyment and participation of the local community were the key drivers of events, now there is criticism that festivals are often imposed on communities such as in the case of Kangaroo Island Surf Music Festival which created opposition from community members who felt they had not been consulted in the event planning stage (Higgins-Desbiolles, 2018). Hence, a key factor necessary for enhancing destination community social capital includes effective local resident engagement in tourism planning (Moscardo et al., 2017). Recent critical approaches to events tourism also focus on the need to explore the power dynamics between a festival and the local community (Dredge and Jamal, 2015), in order to provide a more critical deconstruction of the political and economic structures that shape tourism events' policy and planning.

Adding to the already blossoming area of critical events tourism literature, this book aims to fill a gap in knowledge by furthering the discourse on critical events tourism research and as such is likely to be of interest to events planners and event tourism academics throughout the world. This book seeks to further explore the concept of events tourism with a critical focus, and examine its importance from different economic, technological, industry, policy and socio-economic perspectives that will highlight critical issues, showcase best practice examples, raise managerial and political implications, formulate conceptual frameworks and advance understanding in this important area of tourism research.

Our book structure

The contributing chapters are grouped under four broad categories: (1) fundamental concepts, (2) contemporary perspectives, (3) regeneration, displacement and planning frameworks and (4) the future of events tourism. These reflect a broad range of events tourism outcomes that we propose require critical reflection.

In Part I Fundamental concepts, the editors Fiona Bakas, Willem J. L. Coetzee and Violet V. Cuffy present an overview of events tourism and the core issues on which the book is underpinned in this chapter.

Part II broadly covers the topic of contemporary perspectives. First, David Strafford expands on the concept of Destination Shopping Centres (DSC) as "mini cities" increasingly using events to attract "tourists" from further afield, encouraging them to arrive earlier, stay longer and leave later. In Chapter 2, according to Strafford, there is a shift from a core offering of simply somewhere to shop towards more of a leisure facility, with events employed as a key part of this marketing strategy. Focus is on events tourism in a contemporary sense, with non-traditional venues, using non-traditional events, to attract non-traditional tourists. As the pressure of the bricks and mortar retail sector grows, the more wide-ranging, entertaining, experiential and engaging the events will become. This chapter highlights the changing shopping centre market, why consumers are demanding all-in-one experiences and how DSC managers are reacting to this demand through an increasingly strategic and creative approach to their event portfolios.

Birgit Muskat and Judith Mair, in Chapter 3, analyse job experiences and they provide insight into managing event staff heterogeneity. Their chapter offers a deeper understanding of events management by adding an organisational behavioural perspective. The insights in this chapter are relevant to event managers, specifically in relation to staff job experiences and job satisfaction management. Here, staff hold a central position in event management; this chapter synthesises the unique organisational aspects of events and theoretically explores how the event workforce, consisting of permanent employees, casual staff and unpaid volunteers, experience job satisfaction. Although this chapter focuses mainly on major events which attract tourists to a destination and which are large enough to have a complement of full-time permanent staff, elements of arguments in this chapter will also have relevance to smaller events, even those entirely organised and staffed by volunteers.

Estela Marine-Roig, Eva Martin-Fuentes and Natalia Daries-Ramón's chapter follows exploring user-generated events (UGEs) in tourism through social media in Chapter 4. They argue that social media have revolutionised travel and tourism, and special attention has been given to tourist user-generated content (UGC) and the formation of online communities; however, little attention has been paid to tourist events entirely generated by users through social media. The aim of this chapter is to define and characterise the phenomenon of tourism UGEs through social media around users' new role of empowerment and to highlight their interest in tourism organisations. They provide a quantitative framework approach to store, analyse and compare events in relation to social media, which they complement with qualitative observant participation at the events. A mixed-method approach is applied to the analysis of three Instagram meetups organised by a specific online community at three Pyrenean Ski resorts. Results confirm the UGEs' great capacity for image dissemination and attraction, and outline the differential characteristics resulting from this new user empowerment.

Chapter 5 "Undocumented! Small events of rural hinterlands, South Africa" by Unathi Sonwabile Henama provides some insights into events on the margins. He argues that tourism has grown "in leaps and bounds" since the democratic transition of South Africa in 1994, which heralded the end of apartheid in South Africa. The growth of tourism has been so spectacular that today it is regarded as the new gold, South Africa's number one export. The growth of tourism has also been associated with the growth of special interest tourism that has seen the emergence of sports tourism, events tourism, religious tourism and health tourism. For instance, South Africa has hosted numerous sporting events such as the 1995 Rugby World Cup and 2010 FIFA World Cup, as well as several cultural festivals. The plethora of academic gaze has been primarily urban with scant attention given to less glamorous rural events. Key is the celebration of Ramadan in Bo Kaap, a religious Muslim orientated event, whose legacy resulted in the declaration of heritage protection for Bo Kaap. Then, Community-Based Tourism at Komjekejeke, an annual celebration of Ndebele culture. Here the legacy involves upgrading of infrastructure and creation of an enabling environment to consume Ndebele culture.

Keeping with the rural theme in Chapter 6, Burcin Kalabay Hatipoglu, Onno Anıl, Saadet Memiş and Dilan Şahin describe the impact of a food-based event on entrepreneurship development in rural areas. An Earth Market in the Turkish town of Şile, near Istanbul, attracts visitors from nearby regions and food lovers from Istanbul. The authors adapted the empowerment framework devised by Scheyvens (1999) as a tool for answering our research question. The empirical findings demonstrate that the event has empowered the participants psychologically, socially and politically, though tourism businesses are empowered to a lesser extent in all dimensions. From the research, the framework for supporting rural entrepreneurship through a planned event is proposed.

In Chapter 7 of this section, Lucia Cicero explores the concept of wanderlust from the perspective of Motorcycle Fests. This area has evolved over the decades allowing for investigation both as a factor influencing tourism motives by attendees and as a criterion for segmenting the audience of motorcycle events from a marketing standpoint. Findings show that the will to travel is tightly connected with the tourism motivation of motorcycle event participants. Wanderlust participants focus on fun and freedom associated with this event genre as primary motives for their involvement.

Part III which highlights regeneration, displacement and planning frameworks follows with Chapter 8, by Abrar Faisal, Julia N. Albrecht and Willem J. L. Coetzee exploring the earthquake aftermath and transformation of the conference industry in Christchurch, New Zealand. This chapter examines the period during the earthquake aftermath focusing on the recovery trajectory of a post-quake conference industry. Industry insights offer a comprehensive understanding of the critical challenges and opportunities in a post-disaster environment. The first section draws on urban regeneration and conference industry literature frameworks, concerns of an altered city alongside government's responses and legislation. Finally, the conference industry is examined as a potential regenerator in a post-disaster city.

Chapter 9 by Jérôme Massiani argues that Expo 2015 visitors' expenditures had significant benefit for Italy, though below ex ante expectations and at the cost of reduced economic activity in other regions. Jérôme highlights that many evaluations of the impacts of mega-events are overestimated and do not consider the substitution effect. Based on a standardised survey, the added component of visitors' expenditures is set aside and more realistically the impact evaluation is proposed. The results show that almost all visits from Italians are "substitutive" of other presence in Italy, while only half of foreigners' visits are surplus for the country. Arguably, the impact on the case study is that Lombardy is larger than the rest of on the whole country, indicating that other regions were negatively impacted by the event. This mainly relates to the expenditures of non-Lombard Italians, as a huge share of visitors is shown to shift expenditures from other regions at the benefit of Lombardy. Results are contingent on assumptions regarding the number of days spent in Lombardy/rest of Italy; a lower magnitude in the results is reflected when more conservative assumptions are made.

From a different perspective, James Malitoni Chilembwe highlights and discusses critical factors in religious events tourism management in Chapter 10. According to the author, increased demand arises from an increased population globally, leading to more effective event management. He contends that while holding events is good for the hosting country resulting in social and economic gain, there are a cloud of associated negative issues such as environmental degradation. In addition he presents several fundamental challenges requiring attention from key stakeholders. This chapter contends that with the continued growing population around the globe and corresponding increasing demand for event participation around the world, negative trends demand better management of intercontinental and regional events.

Bidding for cultural events is explored through the lenses of city re-branding, social discontent by Daniel Barrera-Fernandez, Miguel Anxo Rodriguez-Gonzalez and Marco Hernández-Escampa in Chapter 11.

The growth of MICE (Meetings, Incentives, Conferences and Exhibitions) tourism development in Ethiopia is an interesting development. In Chapter 12, Gebeyaw Ambelu and Brent Lovelock analyse the current position of and challenges for the MICE tourism development in the context of Addis Ababa, Ethiopia. This chapter is prompted by the relative paucity of attention given to the challenges of advancing the MICE sector from the developing world perspective. As one of the more rapidly growing market segments of the tourist industry, and according to the International Meeting Statistics of the Union of International Associations (UIA), in 2016, there were 468,700 international meetings held worldwide (UIA, 2017). The authors argue that it is now common to see local and national tourism strategies that position events as central components for destinations, a form of attraction, as development catalysts, and for destination branding and image-making.

The fourth and final part of the book covers the future of events tourism in three chapters. Silvia Grandi and Chiara Bernasconi in Chapter 13 address the work of artists in reshaping metropolitan cities via creative tourism.

Daniel Wright presents to us "Counterculture and future of music festivals and events" in Chapter 14. The chapter considers the future of the music festival and events industry, offering discussions surrounding the quest for uniqueness in a standardised commercial environment, taking a counterculture perspective. Gathering knowledge with which to identify countercultural movements and how these could be central to gaining a competitive advantage for future businesses is crucial. The chapter explores culture, society and counterculture as an engine operating in the margins of society, before considering the impact of the internet and social media. Importantly, attention is also on how big data analytics offer an opportunity for festival and event organisers to recognise developing counterculture movements and how these have the potential to become profitable within the mainstream consumer environment.

In the final chapter, Violet V. Cuffy and Bipithalal Balakrishnan Nair offer insight into future debates for the 21st century. This closing chapter highlights the core shifts taking place in the subsector. Ten key drivers are identified as forces of change, ranging from the desire for authentic expenses, power dynamics, augmented reality and robotics based on perspectives of Yeomen et al. (2013). Other key developments are the impact of the changing customer profile and expectations in the area of events and festivals. In conclusion, they highlight that the necessity of continuous training of educators will be evermore paramount as the sector advances.

References

Chalip, L. (2006) 'Towards social leverage of sport events', *Journal of Sport & Tourism*, 11(2), pp. 109–127.

Costa, C. et al. (2017) 'Gender, flexibility and the "ideal tourism worker"', *Annals of Tourism Research*, 64, pp. 64–75. doi: 10.1016/j.annals.2017.03.002.

Dredge, D. and Jamal, T. (2015) 'Progress in tourism planning and policy: A post-structural perspective on knowledge production', *Tourism Management*, 51, pp. 285–297. doi: 10.1016/j.tourman.2015.06.002.

Getz, D. (2008) 'Event tourism: Definition, evolution, and research', *Tourism Management*, 29(3), pp. 403–428. doi: 10.1016/j.tourman.2007.07.017.

Getz, D. and Page, S. J. (2016a) *Event studies: Theory, research and policy for planned events*. London: Routledge.

Getz, D. and Page, S. J. (2016b) 'Progress and prospects for event tourism research', *Tourism Management*, 52 (Supplement C), pp. 593–631. doi: 10.1016/j.tourman.2015.03.007.

Hannam, K., Mostafanezhad, M. and Rickly, J. (2016) *Event mobilities: Politics, place and performance*. London: Routledge.

Higgins-Desbiolles, F. (2018) 'Event tourism and event imposition: A critical case study from Kangaroo Island, South Australia', *Tourism Management*, 64, pp. 73–86. doi: 10.1016/j.tourman.2017.08.002.

Kim, W. et al. (2015) 'Evaluating the perceived social impacts of hosting large-scale sport tourism events: Scale development and validation', *Tourism Management*, 48, pp. 21–32.

Lamond, I. R. and Platt, L. (2016) *Critical event studies: Approaches to research*. London: Springer.

Mair, J. and Whitford, M. (2013) 'An exploration of events research: Event topics, themes and emerging trends', *International Journal of Event and Festival Management*, 4(1), pp. 6–30. doi: 10.1108/17582951311307485.

Moscardo, G. et al. (2017) 'Linking tourism to social capital in destination communities', *Journal of Destination Marketing & Management*, 6(4), pp. 286–295. doi: 10.1016/j.jdmm.2017.10.001.

Pernecky, T. (2013) 'Events, society, and sustainability: Five propositions', in T. Pernecky and Luck, M. (eds) *Events, society and sustainability: Critical and contemporary approaches*. London: Routledge, pp. 33–47.

Raj, R. and Griffin, K. A. (2015) *Religious tourism and pilgrimage management: An international perspective*. London: CABI.

Ren, C., Pritchard, A. and Morgan, N. (2010) 'Constructing tourism research: A critical inquiry', *Annals of Tourism Research*, 37, pp. 885–904.

Scheyvens, R. (1999). 'Ecotourism and the empowerment of local communities', *Tourism Management*, 20(2), 245–249.

Spracklen, K. and Lamond, I. (2016) *Critical event studies*. London: Routledge. doi: 10.4324/9781315690414.

Stevenson, N. (2016) 'Local festivals, social capital and sustainable destination development: Experiences in East London', *Journal of Sustainable Tourism*, 24(7), pp. 990–1006.

Tribe, J. (2008) 'Tourism: A critical business', *Journal of Travel Research*, 46, pp. 245–255.

Yeoman, I. (2013) 'A futurists thoughts on consumer trends shaping future festivals and events', *International Journal of Event and Festival Management*, 4(3), 249–260.

Ziakas, V. (2016) 'Fostering the social utility of events: An integrative framework for the strategic use of events in community development', *Current Issues in Tourism*, 19(11), pp. 1136–1157. doi: 10.1080/13683500.2013.849664.

Žižek, S. (2014) *Event: Philosophy in transit*. London: Penguin.

Part II

Contemporary perspectives

2 Events as Visitor Attractions in Destination Shopping Centres

David Strafford

Introduction

Shopping centres can no longer rely on shopping alone to attract custom-ers (Howard, 2007). The international retail sector is facing mounting pres-sure from online shopping, forcing a notable shift among large, Destination Shopping Centres (DSCs) to invest in the creation of unmissable experi-ences, using events as a key part of their strategy (Backstrom and Johans-son, 2006; Getz, 2007). DSCs, seen as artificial mini-cities, endure increased competition and have therefore augmented their offering, encouraging their 'guests' to spend time, as well as money, in their spaces (El-Adly, 2007; Howard, 2007).

This chapter will expand on the idea that these DSCs are increasingly us-ing events to attract 'tourists' from further afield, encouraging them to arrive earlier, stay longer and leave later. There is a move away from their core offer-ing of simply being somewhere to shop towards more of a leisure facility, and the use of events is a key part of this marketing strategy. This is event tourism in a contemporary sense, with non-traditional venues, using non-traditional events, to attract non-traditional tourists; however, the shift is real and as the pressure of the bricks and mortar retail sector grows, the more wide-ranging, entertaining, experiential and engaging the events will become.

The range of events that sit within this postmodern retail environment is examined. Further investigation is conducted to illustrate how a vibrant portfolio is examined, with the Dubai Shopping Festival as the prominent focus. The changing nature of centres in Hong Kong is significant – two international city-states that embody the commercialisation and commod-ification of space and ultimately blur boundaries of retail, events, tourism and hospitality. Other UK case studies include the DSCs at Meadowhall, Trafford Centre and Centre: MK.

This chapter will first focus on the changing shopping centre market, then why consumers are demanding an all-in-one experience (Arnold et al., 2005) and finally and more extensively, how DSC managers are reacting to this demand through an increasingly strategic and creative approach to their event portfolios.

The changing face of retail

In China, visitors can witness the 'world's saddest polar bear'. This animal resides not in a zoo, in the wild or even in a sanctuary, but in a cage inside the Grandview Shopping Centre in Guangzhou (Molloy, 2016). The polar bear is subjected to numerous selfies from shoppers and has recently been temporarily rehoused after he showed signs of mental distress (Molloy, 2016). This story is not alone – in fact far from it. In the Mall of Emirates, Dubai, shoppers can take a break from their retail activities with a quick diversion to the indoor ski slope, complete with real penguins (Mall of the Emirates, 2017). In West Edmonton Mall in Canada, there are daily sea lion shows amidst a giant indoor water park (Thomassen et al., 2006). Indeed, this Retail Centre is of particular interest owing to its tagline of "the greatest indoor show in the world", a hyperbolic marketing message straight from the world of entertainment and events, playing on the overwhelming sense of the unmissable (Tressider and Hirst, 2012). These examples indicate a growing pattern from Retail Centres to look beyond simply offering consumers a place to shop (Jones, 1999). This fusion of entertainment, leisure, events and retail under one roof is a growing one, with Retail Centres adapting to changing shifts in consumer demands for ease and convenience, but with added wow factor (Howard, 2007; Michelli, 2007).

Indeed Dubai is one destination that has embraced these concepts particularly strongly. Dubai itself is arguably an entire region whose tourism industry is largely built on retail; and yet now there is an increased focus towards the Visitor Attractions industry (Mehta et al., 2014) and the ensuing and inevitable merging of the two worlds. There is currently colossal investment in the Visitor Attraction industry in Dubai, with over $350 billion promised by 2020 (Picsolve, 2015) as it seeks to expand its inward tourism, using both crafted and artificial stimuli. The perceived lack of ancient cultural heritage in Dubai (as opposed to say Egypt or Greece) means new avenues of generating tourism are pursued. It is the ultimate living embodiment of this postmodern ground breaking hybrid approach, fusing entertainment, retail marketing, tourism and events in commercial settings. Dubai invests in its infrastructure and tourism like no other city and is the ultimate example in consumer retail experience; indeed its tourism industry is founded upon it. The January 'Shopping Festival' in Dubai is the pinnacle of this – a perfect example of these fields merging. In fact, the Shopping Festival's marketing campaign is of interest since it captures the true embodiment of the current trend to appropriate the word 'Festival' by sectors other than events. The Dubai Shopping Festival lasts for over a month (starting at the beginning of January) and the marketing borrows imagery and semiotics from the events world (Tressider and Hirst, 2012). Questions raised by merging concepts of 'retailtainment' and Visitor Attractions are echoed by Retail Centres everywhere. The questions remain as to where Retail Centres fit into modern day life and where they fit into the sociological perspective of human existence in a postmodern world, of which Dubai is the most concentrated form. They are still relevant, existing alongside online and digital shopping, providing a central hub for family life, community life and social interactions. Howard (2007) argues that modern time pressures now mean that family time is shopping time, and vice versa.

The implications of this, for retailers, are that they need to understand the motivations of shoppers in greater depth. Shoppers can be either interested in efficient, functional shopping (utilitarian) or they can be more interested in pleasurable, leisurely shopping (hedonic), taking time over browsing and enjoying the act of shopping itself (Lombart, 2004; Howard, 2007). Indeed, hedonic shopping from a sociocultural perspective involves the consumer moving around Retail Centres, enjoying the environment and deriving pleasure from being there, rather than doing any specific purchase or trans-action (Backstrom, 2011). The act of browsing, or the shopping trip itself, is the leisure activity and the source of pleasure for some consumers – consider those who make the trip to Oxford Street in London for example (Lombart, 2004). The buying of goods and services is incidental, not fundamental, to the social shopper's experience and the social interaction is the focal activity, with any purchasing of items secondary in terms of hedonic value (Backstrom, 2011). The real value, for social shoppers, is spending time with loved ones, whether that be friends or family, mother and daughter trying on clothes for instance (Howard, 2007; Backstrom, 2011). Jones (1999) agrees, suggesting that the social aspects of shopping are more important to some consumers enjoy not being rushed, but taking time to enjoy the shopping experience derived from spending time with friends and family, rather than from the act of purchasing itself.

Borges et al. (2010) go further by suggesting that shopping is now a social activity, enhanced by companions helping with decision-making; and their study found that consumers who shop in groups tend to spend more time in centres and visit more shops, making more purchases in value of sales. Recreational shopping is now the norm, with consumers essentially brows-ing without any specific intent to acquire items; the pursuit of pleasure and stimulation of the senses is the leisure activity in question (Backstrom, 2011). Some consumers enjoy not being rushed and taking their time with shopping experience, savouring it, according to Jones (2010). Indeed, Howard (2007) agrees by suggesting that the shopping trip itself is the whole point of the activity, not necessarily the act of buying goods and consumption. Purchas-ing items is only part of the activity. Consumers seek approval from friends and family before making purchases, according to Borges et al. (2010), to which Underhill (1999) agrees, suggesting that shoppers who are with friends spend longer in stores and are more likely to purchase, than shoppers who are alone. Ultimately this means the postmodern consumer is finding a sense of community and identity in the act of shopping and therefore in DSCs. Consumers prefer stores and Centres where they believe they will find simi-lar communities, providing a sense of self-worth and identification through a shared consumption of brands (Borges et al., 2010). Consumers are seeking identification and belonging to communities by visiting Retail Centres, encouraging a sense of finding themselves and increasing hedonic value. For those customers who do identify with the Retail Centre, Borges et al. (2010) argue, this can lead to increased expenditure of time and money – the consumer feels part of community and is, therefore, more comfortable.

Retailers should facilitate playfulness in-store, creating memorable customer experiences in order to elicit hedonic values of fantasy, fun, satisfaction and even escapism (Backstrom and Johansson, 2006). Woodruffe-Burton et al. (2001) refer to Retail Centres as 'fantasy palaces'. There is a drive to move away from functional shopping experiences and move towards a more immersive, enjoyable and social shopping experience for the consumer (Backstrom and Johansson, 2006). There is a desire for Retail Centres to provide engaging experiences for consumers, who are involved and excited by the time spent in the centre, encouraging satisfaction, loyalty, increased time and money spent, leading to repeat visits. For Davis and Hodges (2012), the value for consumers is channelled through utilitarian means (ease and convenience), hedonic dimensions (playfulness, enjoyment and fun) and social value (self-esteem and status). It appears that Retail Centres need to offer all three dimensions to consumers to compete with the increasing threat from online shopping. Customers expect a greater value experience, if they are making the effort of actually visiting a shopping centre.

The generation of *shopping value* (a holistic concept that combines all elements of the shopping experience) is fundamental to Retail Centres' strategy of retaining elements of competitive advantage (Davis and Hodges, 2012), the central key being the motivator for consumers to actually visit the bricks and mortar centre instead of shopping online. Arguably, hedonic value and social value are difficult to generate from internet shopping, which is precisely what sets Retail Centres apart. Retail Centres need to embrace experiential consumption and increase the value of the physical trip to the store. It is an intense source of competitive advantage for retailers, to create memorable and engaging experiences, with the use of special events increasingly central to this strategy (Davis and Hodges, 2012). Retail Centres need to appeal to consumer demands for functional utilitarian shopping experiences, with customers' basic needs being met, but they also need to address the desire for pleasurable, hedonic shopping experiences, indulging in multi-sensory experiences, eliciting feelings of fun and fantasy, surprising and delighting their consumers (Jones et al., 2006).

Therefore, it is no longer enough for retailers to focus on quality of their service, and merely satisfy customers – indeed, customers expect to be satisfied (Arnold et al., 2005). The raised bar, with retailers needing to focus on a loftier goal of delighting and exciting their postmodern customers, goes beyond what is expected. Shopping centres should create incentives for shoppers to feel part of a community, in order to increase hedonic value from

BOX 2.1 INDUSTRY VOICE: ALEX CALEY, MARKETING AND EXPERIENCE MANAGER, MEADOWHALL SHEFFIELD

Meadowhall is a Super-Regional shopping centre, essentially meaning you pull shoppers from quite a way – you are a destination. In terms of our catchment, we've got our primary catchment, which is on our doorstep in Sheffield, but then we've got our secondary area (Doncaster, Barnsley, Rotherham, Chesterfield) and then our tertiary (Leeds, Bradford, Nottingham and so on). Similar Super-Regional Centres would include Trafford Centre, Bluewater, Westfield etc and our customers want more than just retail these days. I think customers have an expectation, particularly now, beyond just bricks and mortar: people discuss online shopping and ask us if it's the death of shopping but all the trends – as much as online is growing – suggests there is still definitely a place for bricks and mortar retail. But it's more a case that people expect more in terms of their experience; whether in terms of being connected to WIFI, animated spaces, ease and convenience etc. People are just more into experiences and so it's only right that, as a landlord, we deliver what people expect. It's not enough to just have shops now, we've got to go over and above. British Land's [who own Meadowhall] strapline is 'places people prefer' and we like to embrace this idea of 'enlivenment' which links to animation, vibrancy. It's about making a difference to a similar space, to animate the space. So, as a landlord, the easy option is just letting the shops do whatever they do, but what is increasingly common is more of a focus on these places in between the retail, so they're the gaps in between those spaces, which we call the Mall. It could be some kind of brand personality activity; for instance putting a piano there and encouraging the consumers to create the enlivenment. Or it could bigger events such as the ones here at Meadowhall including Ladies Night, or Student Lock-In, or Christmas Live – it's critical that you have a mix. It's difficult to do something on a daily basis of course, but we're trying to fill those gaps, either through our connectivity or digitally, or socially, we think it's important to have that personality which fills in the gaps between the retail. It's hard to measure a return on marketing investments, whether it's a TV or a radio campaign or an outdoor mailer or whatever it might be; however with a large student event for example, you can see that from your investment, you have brought in 22,000 students! We can measure their footfall, we can measure their spend – so, all of a sudden, the retailers are happy because they can see the results, and they know their tills rung that night.

shopping, encouraging consumers to feel part of this wider community and spend more time (Howard, 2007; Borges et al., 2010). Indeed, the implication for this is that Retail Centres now appreciate that some shoppers seek to experience malls in other ways other than just shopping – primarily socialising with friends, with no intention of buying goods, just browsing. Consumers will drive past weaker centres to arrive at the 'destination' Retail Centres, in order to maximise their shopping experience (Wakefield and Baker, 1998). The concepts of social shopping and browsing are therefore, central aspects of postmodern consumerism.

Retail Centres as destinations

The recognition that events have impacts and organisations can seek to harness those impacts through a strategic and purposeful event policy. Events are increasingly being used as tools to bring short- and long-term social, cultural and economic impacts to city centres for instance (Morgan et al., 2011) and as such events can provide an effective way of bringing communities together and boosting economic growth through visitor expenditure and tourism (Richards and Palmer, 2010). Major and mega-events are increasingly being used by city councils and event management teams to drive not only economic benefits but also encourage social, political and cultural change (Getz, 2007), since effective marketing and successful events can inherently improve the reputation and brand of the host city (Morgan et al., 2011). A successful events policy can incur wholesale changes to a city's image, according to Richards and Palmer (2010), with Glasgow, Barcelona and Dublin setting recent positive examples (Heeley, 2011). Events can provide long-term branding and tourism benefits by shaping the perceived images of the host city (Getz, 2007) as well as driving footfall, primary and secondary spend and bed nights on the event dates (Richards and Palmer, 2010).

There is a need for all destinations to consider their own competitive advantage (Heeley, 2011), and discover what it is that makes them different from other similar places (Davidson and Rogers, 2006). It is a crowded market (Richards and Palmer, 2010); tourists and event tourists have an understanding of what it is that encapsulates a place's identity, ostensibly what makes it worth visiting (Davidson and Rogers, 2006). Events can act as showcase opportunities for destinations, with urban areas being transformed into stages, and city scapes as backdrops (Richards and Palmer, 2010). The image of London on New Year's Eve is intrinsically linked with the fireworks on the Millenium Eye, and Times Square forms the backdrop and performance area each year for New York's new year celebrations. These cities understand the value of destination marketing through events and the control of the media (Wang and Pizam, 2011), and, arguably, Retail Centres are starting to understand that they should do the same. Public festivals also contribute to the social cohesiveness and the goodwill of the country, leading to increased tourism and higher economic impacts (O'Toole, 2011) and cultural strategies should be consultative, dynamic, performance-driven, long-term and team-oriented (Pike, 2012). Governments (both local and national) can encourage growth in the events industries; they can also invest into arts programmes to fund events to drive event tourism and spending (Tribe, 2011). They can invest in infrastructures, transport links and communications, which are all vital parts of event management (Getz, 2007). Simply put, mega-events are often used to kick-start local economies through urban renewal and infrastructure investment, even the ill-fated and much-maligned Millennium Dome had the social legacy of bringing an underground Tube station to Greenwich, a depressed area of southeast London, and there are now increasing signs of regeneration in this area, with new housing and retail developments (Tribe, 2011). The more expenditure in a destination during an event, the more tax revenue raised and secondary spend on items such as taxis, catering and hotels can increase the vitality of the area (Richards and Palmer, 2010).

Much literature exists on public sector cities and destinations using this strategic event policy to achieve outcomes (Dredge and Whitford, 2011). Events can help cities meet organisational objectives and goals such as a more diverse community, a more accessible city centre, collaborative working and sustainability (O'Toole, 2011). Tourism objectives can also be achieved through events policy at national and regional levels, with increased visitors, productivity and primary and secondary spending (Richards and Palmer, 2010) and tangible and intangible benefits can be gained by a strong events strategy (Pike, 2012) with measurable impacts that links in with the national agenda (Foley et al., 2012). Debatably, events and festivals are employed to achieve social outcomes, attract investment, regenerate areas, build communities and attract tourists (Morgan et al., 2011). A range of authors have focused on these social outcomes as being part of the purposeful event policy, which can be contrived by local or national governance, adding to a sense of inter-city competition. A coherent and choreographed event policy can determine a range of successful strategic outcomes for a city or destination, with tourism, investment and community outcomes amongst the intended impacts (Morgan et al., 2011). A well-planned and well-strategised event policy can be a useful strategic tool for government (Foley et al., 2012). Where once 'unplanned' and 'organic' events occurred in cities, with authentic and natural community-based origins, now encourage a range of manufactured events, bound together in an over-arching destination's portfolio (Ziakas, 2013). Hobswarm and Ranger (1983) called this the 'invention of tradition' with carefully crafted, manufactured events and activity, designed and contrived to enhance the destination image.

Destinations such as shopping centres need to be 'eventful' of course. There are a range of event strategies that they can employ which all have downsides to sit alongside the benefits (Foley et al., 2012): through creating their own from scratch (risky), embracing the existing programme (haphazard) or buying in touring productions (expensive). Cities can of course bid for footloose events, attracting touring events such as the Olympics, the World Cup, the Grand Prix and other sporting/music tourism events – but what of shopping centres? Economic drivers are central, with short-term and long-term tourism and investment gains being sought, but social-cultural and reputational gains are consistent with a successful event policy (Foley et al., 2012). Long-term reputational change is of particular interest, with cities and destinations recognising that their event portfolio tells a story about what kind of destination they want to be (Ziakas, 2013). The different event strategies adopted by New Orleans (fun, party and music), Las Vegas (glamorous, excessive, entertaining), Dubai (luxury, wealth and commerce) show how powerful event policies can be, in determining the marketing image of the destination. With this in mind, it is critical for cities and destinations to embrace event policy as a strategic tool (Gerritsen and van Olderen, 2013) to not only affect real investment and social impacts at ground level but also shape the global narrative of that destination (Urry, 1995) which is critically important in the age of Instagram and Snapchat. These social media sites are the 'shop window' (Hobbs, 2015) for destinations to attract tourists, and what better way to generate engaging social media content than through events and activations.

For Retail Centres, the story is no different. In essence, they are destinations, with customers acting as visitors, attracted to spend time and money at their locations. The marketing and event management teams at Retail Centres are borrowing techniques and strategies from cities, destinations and even Visitor Attractions, to encourage their audience to make more repeat visits and spend more time and money (Litvin, 2007). Ultimately, the marketing strategy of a Retail Centre is stakeholder-driven, with management needing to design and choreograph activity, in order to achieve pre-determined outcomes. For a commercial organisation such as a Retail Centre, these outcomes will include more customers, increased profit, increased dwell time and an expansion of the catchment area. However, other softer objectives will also apply, including improved customer satisfaction, good news media stories, improved sustainability scores leading to an improved reputation and positive brand. Retail Centres must consider branding in the same way as other commercial organisations do (Morgan et al., 2011), but also in the same way that city centres and Visitor Attractions do – critically there is a need for them to be *attractive*. A destination's attractiveness is at the heart of its ability to grow tourism.

Retail event tourism (using events to attract tourists)

Acceptance that a DSC, such as Meadowhall in Sheffield or Westfield Stratford in London, is able to take on the characteristics of a city in terms of destination marketing, eludes to the application of principles of event tourism. Getz (2007) argues that cities can use events as tourist attractions to further the destination marketing and image creation of that place – and

BOX 2.2 INDUSTRY VOICE: NIKKI TANSEY, EVENT MANAGER AT INTU TRAFFORD CENTRE

intu own many shopping centres in the UK, and we have very cool values, being enjoyed, being creative, being different, and as part of our brand we have brand pillars – and one of our brand pillars is 'events with a difference'.

So that's where intu really strives to be different within their shopping centres, by putting on different kinds of events that you can't really get elsewhere, so this is the strive for intu going forward as a brand. And so, intu Trafford Centre obviously fits into that. So in terms of why we do events, we have two things, so intu are very big into the whole customer journey, from a customer service point of view; but the customer service is the transactional part we would say, so for instance selling items, it's about how we sell that item and the experience that is attached to it. So, the event would help people come in, they would see something and that would make their experience a lot better. That's basically the kind of gist of why we do what we do, to make them feel happier when they're here – and the events and entertainment make people feel happier. And if they're happier they will stay longer. Thus we have a marketing strategy and the events fit within that, and our objectives would be to increase dwell times, increase footfall and we use events throughout the calendar year at key points – so we know the key points – for instance, at the minute we've got summer holidays and in the past we have always looked at bringing families in for instance and that's still right, but now we need to look and research further into our key audience, so we do market research. At the minute we're trying to get more of the affluent shoppers, so not just the children who are off on their six weeks holidays, but you've got the evening trade as well, which we really need to maximise and we use events to do that. So, you've got to get your head into all those audiences and what do those audiences want and put on an event for that and that does fit in line with the marketing strategy. We are always trying to place PR at the heart of everything that we do, because we're very conscious that people can come in and see an event and they would have to be in the centre to see it, but we want that message to reach far and wide, so it needs to be of a PR-able nature, whatever we do in Centre, we get into the print media, we get on social media and the message goes far and wide, to raise brand awareness more than anything as well. It's not a standalone event, it all needs to hang together.

indeed this is an accepted view. Events can shape consumers' perceptions of a place both for good (Las Vegas benefits from hosting glitzy and glamorous entertainment events) and bad (Las Vegas also suffered from the 2017 shootings at its country and western music festival). How possible is application of ideas of event tourism to DSCs?

It is critical for places to maintain their overall destination image – and without doubt, events feed into this (Pike, 2012). The range, scope and size of events hosted at a DSC will influence consumer perception of the centre and (either consciously or subconsciously) will influence whether they choose to visit. Some centres will use larger events to attract customers from further away, on the assumption that consumers will drive past weaker stores to attend one-off events. Events such as Christmas Lights Switch On events, or book signings, can act as animators of the centre – driving footfall to the centre, perhaps for the first time (Getz and Page, 2016). The Centres then hope that these visitors decide to first stay for longer than the event (e.g. have lunch after the book signing) and second choose to repeat their visit another time when there are no events on. Centres can use different types of events, to attract different market segments and for this reason, events are extremely useful for marketeers: they are flexible, they are short-term, they are temporary. They are a nimble and often quick response to trying to attract a range of customers. Whilst some Visitor Attractions might primarily focus on one target segment, DSCs need to be open to all, and being able to host a range of events supports this. Inviting a YouTube Influencer for a guest appearance might appeal to a teenage audience, but this can sit happily alongside a book signing from a Radio 4 political show host appealing to older audiences.

These images of the events combine to create an overall impression of the Centre. In the same way as other destinations, marketeers can choreograph a range of different types of events to come together in a particular, desired image of the centre to its consumers (Pike, 2012). It might be that a Centre wants to appeal to a more affluent crowd, and therefore its events would need to focus on luxury lifestyle brands. Alternatively, it might be that it wants to attract more females, and therefore events can focus around more female-oriented brands, products and celebrities. The more sophisticated the event portfolio, the more coherent the marketing strategy will appear, and thus the more successful it is likely to be (Ziakas, 2013). Equally, a strong event portfolio can attract consumers for the two main objectives – to stay for longer than the event and to repeat visit (Ziakas, 2014). This twin approach are the two key factors for events at any DSC – there is a critical need to increase footfall and simultaneously increase dwell time. They want their consumers – or tourists – to come for the event and stay for the shopping and the leisure facilities, crucially also express a desire to return. Events may serve to overcome negative imagery about a centre – perhaps that the Centre is not affluent enough, or busy enough. Events may also appeal to particular tourists – they are indeed powerful marketing tools (Gerritsen and Van Olderen, 2013).

Events can be useful for Centres to improve their infrastructure, in the same way as events have this impact on other places such as host cities for sporting events. A Centre might consider investing in new car park facilities, extra toilet facilities, improved WIFI or even just extra street furniture on the malls, given the extra demand caused by the events. External single phase or three-phase power can be installed to facilitate external events, or perhaps that bit of wasteland next to the car park can be smartened up to make way for the fun fair or the touring food festival, e.g. the installation of a stage in the centre of the mall, to facilitate talks or fashion shows; or perhaps extra lighting or screens can be installed around the Centre, to more easily facilitate impromptu events. The simpler and easier it is for Centres to use events for animation and vibrancy, the more likely they are to take place. As Centres become increasingly competitive, the quantity and quality of their event programme can genuinely be a point of competitive difference, if properly leveraged (Jago et al., 2003; Ziakas, 2014).

A further advantage of using events to manage tourism to DSCs concerns seasonality. Events may serve as campaign launches and to drive footfall to centres on otherwise quiet weekends. The take-off of Christmas is such a critical point in DSC calendars and usually announced through a large event, attracting as many tourists as possible. Event tourism as hosted by shopping centres can focus on attracting the customers they want, when they want, and expert marketeers are increasingly using sophisticated approaches to appeal to different target markets. As data capture around footfall and customer type becomes more intense and knowledge and understanding of consumer habits also increase, DSC marketeers are able to craft and create diverse event programmes. These are designed specifically with the desire to attract customers at varying points in the year. For instance, Meadowhall operates a 'Student Lock-In' event at several points of the calendar on dates when traditionally the centre would be quieter, but also when student loans are freshly released to their bank accounts. New football kit launches help drive footfall in the quieter summer months; children's entertainment shows can be booked for half terms and Easter to capitalise on parents wanting to spend time with children and fill the days. Clever programming can address seasonality issues at Centres, helping to level out the revenue peaks and troughs (Ziakas, 2014).

However, events are not always the answer for shopping centres, in much the same way that they are not the answer for city centres and other places. In the same way events are utilised to attract different market segments; they can also keep other customers away, known as displacement (Wang and Pizam, 2011). Consider the effect that an inner-city urban music festival has on a city centre. The music festival may attract hundreds of thousands of excited teenage tourists from across the region or country – but simultaneously keep away the regular shoppers at department stores. Whilst the bars and clubs benefit, the traditional high street anchor chains suffer. This is the same story for DSCs: at the Christmas Lights Switch On event, the

Centre tends to fill with people for the event, but the traditional customers might have stayed away for fear of it being too crowded. This displacement effect is a classic result of event tourism and it absolutely happens in centres too – some people have a desire to avoid crowds and are displaced from the Centre whilst the events are happening (Wang and Pizam, 2011). Another consideration here is that the crowd for the Lights Switch On might be large, helping to achieve footfall figures and dwell time statistics, but are they realistically going to be also spending money in Boots, M&S or John Lewis during this visit? This can cause disengagement amongst the stakeholders at Centres – namely the retailers (Dredge and Whitford, 2011).

Event portfolios as marketing strategy for DSCs

Events are increasingly becoming central to marketing strategies, which seek to improve attractiveness and it is easy to see why. Events are temporary, requiring little long-term infrastructure investment (Getz, 2007). Events are also transient and can be easily moved, changed and moulded; they are flexible and they are fluid (Berridge, 2007). They add vibrancy and life to sterile environments – and most Retail Centres are new buildings, which need vibrancy adding! Events can add genuine warmth, fun, life and vibrancy to a place that has little or no heritage. They add an element of authenticity. They can give a sterile, bland and boring space a reason for people to visit. The unique aspect of events is that they are unmissable (Tressider and Hirst, 2012); they by their very nature exist for a restricted time, thus giving Retail Centres' customers a reason to return. Arguably, Dubai perfectly encapsulates this phenomenon, by adding retail, entertainment and events to what essentially is a brand new artificial city, giving it vibrancy, life and excitement. They have given people a reason to visit. They have made it attractive. Marketing Managers are excited about events, as they are designable and they are controllable; they generate headlines and they grab attention. They can generate publicity for a space in ways like no other – when the brand new Mall of Qatar opened, the opening party featured live entertainment, stilt-walking trees, shaolin warriors and balloon workshops, in an event they called "a day full of surprises and festivities" (Mall of Qatar, 2017). Finally, events are particularly useful for Marketing Managers of Retail Centres because they allow the marketing to target particular market segments.

Interestingly, now more settled and established as destinations in the mind of the populace Retail Centres accept this strategic marketing approach. Retail Centres of the size of Trafford Centre or Meadowhall are as large as many other town centres in the United Kingdom, and they need to attract more visitors and customers to generate adequate consumption levels. The central ideas behind destination marketing, which previously only really applied to public sector towns, cities, regions and countries, arguably now also apply to Retail Centres. These centres are artificial and manufactured mini-cities, with all the problems that actual town centres have – transport, communication, security, utilities. They are, of course, free entry to the public. Most Retail Centres do not charge an entrance fee, and they have public transport, open car parks (though not always free to park just as city centres). Unlike Visitor Attractions such as Alton Towers or Flamingo Land, Retail Centres are free entry public access, which of course brings with it all the other social-economic problems experienced within towns and cities. The desire to attract increasingly affluent customers must balance with the need to have open public access for most of the daytime. Retail Centres have their own laws, police, security and protection – as privately owned public space, a postmodern oxymoron that is in itself an interesting contradiction.

However, Retail Centres can learn from city centres and destinations – in terms of how they can use events to generate tourism and increase footfall (Morgan et al., 2011). Entertaining event experiences can not only attract more customers but also retain the current ones. Retail Centres now see themselves as a day out for the family – the consumption of family leisure time can now be under one roof, for your convenience. They can also learn from Visitor Attractions, who according to Swarbrooke (1999) need to consider elements such as reputation, branding, customer service, quality and infrastructure as part of the marketing mix. For instance, Disney has focused on customer service which is its key source of competitive advantage – and this is well recognised at the higher end Retail Centres such as Westfield and Bluewater, with the appearance, competence, training and attitude of staff being critically important to the marketing strategy. Another aspect of the Marketing Mix that Swarbrooke (1999) focuses on for Visitor Attractions is the overall feel of pricing – some attractions can adopt a high pricing strategy across all aspects of food, beverage, entertainment and travel; and in a similar way, Retail Centres can too. The portfolio of retail outlets in the centre can determine the overall feel of how expensive a Retail Centre is. A centre with a Harvey Nichols and a John Lewis can afford to be a more expensive commercial proposition than ones without.

Ultimately, Retail Centres blur the lines between public and private space. They are destinations in their customer's minds, and they are a place where some consumers like to spend time. Retail Centres are encouraging longer dwell times by customers, with the addition of night time entertainment, extensive food and drink offerings, and diversified product offerings. To this end, they are now adapting and becoming more like city centres and destinations, with a free and open entrance policy attracting a wide demographic. According to Swarbrooke (1999), destinations are usually larger areas of interest and may indeed consist of a range of Visitor Attractions (Leask, 2010), which in some ways mirrors the Retail Centre experience, in that they are large destinations in themselves, containing within a variety of different attractors (shops, restaurants, entertainment and leisure). However, they also need to look beyond the city centre model and analyse the Visitor Attractions industry; these facilities are privately owned, they are seasonal and they are increasingly designed to retain customers for as long as possible.

Pine and Gilmore (1999) asked of retailers what they would do differently if they charged an entrance fee to their shops and this is where Retail Centres are at currently. If the Trafford Centre had an entrance fee, what would need to change? The central concern here is the visitor experience.

As there are reasons on the supply side why Retail Centres would want to host events as part of their marketing strategy, there are also reasons on the demand side. Of course, Marketing Managers and event organisers seeking to grab attention in high footfall locations cannot dismiss the intensity, convenience and widespread appeal of a shopping centre. There is also a new trend for unusual locations: the power of surprising your audience and challenging them in new and exciting locations (Whitfield, 2009; Higgins, 2017). In the same way that conference organisers now turn to Visitor Attractions to excite and engage their audiences (Whitfield, 2009), many event organisers are seeking to place their touring event productions and engaging marketing activities in Retail Centres. The competitiveness amongst venue providers means diversification of services is a natural and indeed necessary step.

Considering a Retail Centre might operate a variety of events, in order to attract a variety of customers arguably, this policy approach is consistent with public sector event portfolio strategy (Ziakas, 2013). Where a city centre for example might consider designing a varied event policy to include all aspects of the community, attract tourism, generate tourism and infrastructure and build the city's marketing, so might a Retail Centre. In fact, it is this approach to city branding that is so central to the shift in Retail Centres' event marketing strategy. A purpose-built event programme, which appeals to key target audience demographics, can do much to improve the attractiveness of the place and enhance its destination status (Morgan et al., 2011). This also drives the narrative of the branding, allowing Marketing Managers to tell a particular story about their centre. One week it might be a story of new facilities, one week a story of cheap discounts, one week it might be about a new outlet opening. Retail Centres oversee a large community of stakeholders and a carefully choreographed events programme can do much to achieve objectives (Sharples et al., 2014). Meadowhall in Sheffield for instance operates a regular student lock in event aimed at the 16–25 year old market, a regular Ladies night aimed at female shoppers, regular children's events aimed at the family market and then of course wider discount-driven sales aimed at bargain hunters (Meadowhall, 2017). A purposeful event policy can assist Marketing Managers with targeting their customer base, giving flexible and strategic tools to achieve objectives (Foley et al., 2012). The combination of these individual events must, however, add up to tell one unique and single story about the destination – the policy must be coherent; otherwise the events will seem incongruous and disingenuous, and ultimately the holistic policy will be less successful as a result. How Retail Centres design their event programmes is a complex example of stakeholder management, with the need for community-based objectives to balance with financial ones, but ultimately, unless the event can naturally fit into and add to the Centre's overall core brand, it should not belong to the Centre (Morgan et al., 2011).

Indeed this is an interesting difference between Retail Centres and cities as event hosts. Many events that happen in cities are driven from the ground upwards, by the community itself (2013). As such, there is a greater acceptance in cities that the event programme will be more haphazard and sporadic. Community-driven events might not fit under the overall desired city branding (think about an illegal rave in a city centre), whereas Retail Centres have the opportunity to shape and control their complete event programme. Whilst there is a sense of community at Retail Centres, perhaps currently this community is more of a consumer of the space, rather than a co-creator of activity. Retail Centres' Marketing Managers currently dictate their event programmes and whilst they might not necessarily create all the events themselves (proprietary events) they can determine exactly which of the externally organised (non-proprietary events) are staged. It is this sense of control of the destination image that makes Retail Centres a particularly interesting phenomenon to study in postmodern society, especially given the intriguing issue of it between privately owned yet public space.

BOX 2.3 INDUSTRY VOICE: KIM PRIEST, MARKETING MANAGER AT CENTRE: MK IN MILTON KEYNES

Shopping is definitely changing, so if you look at how rapidly retail has changed, especially in the face of the internet, we have to make ourselves relevant to today's customers more and more. I mean why would people bother coming to a shopping centre? This one particularly has parking issues because parking is very expensive, whereas some shopping centres are free, our car parking is run by the council (the majority of it is). [So why would people come here, spend a fortune on parking, have all the hassle of driving? So, I think the environmental footprint of coming here whether it's public transport is impacted – or they can just sit at home on their laptop and order whatever they like and it can even arrive the same day these days. So why would they even worry about that?] So I think we've got to start thinking about the way people shop in different ways, so you've got to give people reasons to come here, so that they want to come to the centre, rather than they need to come to the centre. The retail mix is important, so we do have to have the big retailers, the big retailers will always attract - and there are still quite a lot of people that like that experience of shopping. The fact we've got John Lewis at one end, with Marks and Spencer at the other, we've got Primark, and Next, and Boots. They are, for this centre, what we call Anchor stores. But what else people want when they come to a centre now is more experiential. We have to have fantastic food and drink facilities, lots of dining options. We have to have shops and a retail offer that's slightly different to what you might find on the normal high street – so for example we've got a Mercedes pop-up store down there at the moment. We've just opened a brand new opticians, the likes of which you can only get in the West End of London. So there is that, so they expect a whole day out, the whole shopping experience becomes a day out. The dwell time in the centre is getting more and more and more. We know that people are coming to us from further away so they are making more of a destination of it. We show the cricket for instance and we have a beach in the centre! That's packed with a helter-skelter, punch and judy show, and it's massive for us. Every year we put in the beach and tons and tons of sand. We've got two very key spaces in our shopping centre, again which makes us quite unique, because not all shopping centres have got the space – a huge hall outside John Lewis and it's a huge space, a really big space called Middleton Hall and that is almost like the town square. So it's like the big community space and we put events on there, we have fashion shows twice a year, we have the beach every summer, which is free to enter and it's full of clean, sterilised sand; we've got a fairground ride there, we've got bungees, a helter-skelter, face-painting, we've got a stall selling buckets and spades, candyfloss and the like. People come here early, 9am with their kids and they stay on the beach all day, so they really do make a big thing about it. Screen's Court for the summer (which is Queen's Court), we've transformed that area outside, which is the second space we've got in the centre which is significant. So we've got an indoor hall and an outdoor square and we always like to make sure there is something going on.

A proposed event portfolio model for DSC marketing strategy

We can therefore propose a strategic marketing model of event portfolio for DSCs, which comprises five event types.

The first type of events can be classified as 'high impact' or 'Attractor Events': marquee events, which genuinely attract new and different customer footfall to the DSC. Often this involves bringing a brand in, or a celebrity or a large Visitor Attraction and that attraction is exclusive to the DSC. So for instance the Gruffalo is coming, Wallace and Gromit are coming, which gives customers that reason to visit. The common example here being Christmas Lights Switch-On to announce that 'Christmas has arrived'; however, using events such as The Gruffalo Experience or In The Night Garden Live, to position DSCs as tourist destinations in their own right, means sizable investments are needed more frequently to consistently boost footfall. Moreover, maintaining the authenticity and quality of the Gruffalo event experience and maximising photographic opportunities are critical for achieving the desired objectives.

A second type of events, i.e. 'Medium' or 'Disruptor Events', such as a summer beach-themed festival at centres like Centre: MK or a seasonal Fashion Show at the Ridings in Wakefield, is very much intended as experience enrichment events, rather than 'attraction'. These events disrupt the customer's journey, animating the experience and therein affecting dwell time and customer mood. Customers may decide to stay longer at a DSC because of their journey's disruption.

A third type of events, i.e. 'Enhancer Events', is similar, but smaller, and more habitual, with examples ranging from a piano on the mall to a branded photobooth, beauty makeovers or a treasure hunt, akin to the 'kids club' concepts or photo opportunities. Again the intention is that 'when people are happier, they spend more' creating a persuasive perception that the centre is somewhere where there is always something happening and thus it becomes a place to spend time. In all of the earlier cases, centres discussed the challenge of balancing the requisite short-term footfall, dwell time and sales outcomes, required by tenants, with the longer-term and more subtle imperative to articulate a coherent brand personality that will create engaging experiences. Soft versus quantifiable outcomes were also a key challenge for DSCs as they seek to move away from more hackneyed events and exert a coherent and persuasive brand personality through more immersive experiences. Arguably, all events at DSCs should support brand positioning.

A fourth type of events, i.e. 'Commercial Events', involves selling space for third party use/activation, typically by large commercial brands. Marketing or event executives do not typically make these decisions, but instead by the DSC commercial team often with an exclusively short-term, tactical, operational motive; so there's a purely financial focus to bring events in to impact the DSC bottom line. However, this does not necessarily have benefits for tenants or long-term resonance and is often inconsistent with the overall DSC brand.

The fifth and final category of events, 'Community Events', is CSR-driven events; typically charity or community focused as often DSCs like to be good neighbours. However, the brand-building imperative is still prominent with many charities wanting to operate small-scale fundraising opportunities on the malls.

An investigation into events within DSCs has revealed a prolific and advantageous, but also nascent and intricate, relationship. Existing literature and the voice of respondents are similarly conspicuous in revealing how the changing face of shopping centres underpins a growing, and multifaceted, role for events. In seeking to give people a persuasive reason to not only (re)visit the centres but also 'linger for longer' (Europe Real Estate, 2013), the strategic use of events, alongside permanent 'anchor' leisure attractions, spearheads a sophisticated DSC experience strategy.

Summary

It is clear therefore that shopping centres need to embrace more than just shops to survive, and subsequently compete. They have expanded beyond retail outlets to becoming 'places' in their own right – destinations for tourists to travel distances to. They are adopting the same principles as city centres and Visitor Attractions in that they desire increased footfall and dwell time. They want more people and they want them to stay longer, and events are a key tool at the disposal of a sophisticated and integrated marketing strategy. Events are flexible, they are temporary and targeted, and ultimately, customers now expect some form of entertainment, enlivenment and vibrancy during their shopping visit. Anything else would just be boring.

References

Arnold, M.J., Reynolds, K., Ponder, N. and Lueg, J. (2005), "Customer delight in a retail context: investigating delightful and terrible shopping experiences", *Journal of Business Research*, Vol. 58, No. 8, pp. 1132–1145.

Bäckström, K. (2011), "Shopping as leisure: an exploration of manifoldness and dynamics in consumers shopping experiences", *Journal of Retailing and Consumer Services*, Vol. 18, No. 3, pp. 200–209.

Bäckström, K. and Johansson, U. (2006), "Creating and consuming experiences in retail store environments: comparing retailer and consumer perspectives", *Journal of Retailing and Consumer Services*, Vol. 13, No. 6, pp. 417–430.

Berridge, G. (2007), *Event Design and Experience*. Elsevier, London.

Borges, A., Chebat, J. and Babin, B.J. (2010), "Does a companion always enhance the shopping experience?", *Journal of Retailing and Consumer Services*, Vol. 17, No. 4, pp. 294–299.

Davidson, R. and Rogers, T. (2006), *Marketing Destinations and Venues for Conferences, Conventions and Business Events*. Butterworth-Heinemann, Oxford.

Davis, L. and Hodges, N. (2012), "Consumer shopping value: an investigation of shopping trip value, in-store value and retail format", *Journal of Retailing and Consumer Services*, Vol. 19, pp. 229–239.

Dredge, D. and Whitford, M. (2011), "Event tourism governance and the public sphere", *Journal of Sustainable Tourism*, Vol. 19, No. 4–5, pp. 479–499.

El-Adly, M.I. (2007), "Shopping malls attractiveness: a segmentation approach", *International Journal of Retail and Distribution Management*, Vol. 35, No. 11, pp. 936–950.

Europe Real Estate (2013), "Entertainment in shopping centers: maximizing the shopping experience" [online] http://europe-re.com/entertainment-in-shopping-centers-maximizing-shopping-experience/42425

Foley, M., McGillivray, D. and McPherson, G. (2012), *Event Policy: From Theory to Strategy*. Routledge, Abingdon.

Gerritsen, D. and Van Olderen, R. (2013), *Events as a Strategic Marketing Tool*. CABI, Oxfordshire.

Getz, D. (2007), *Event Studies: Theory, Research and Policy for Planned Events*. Elsevier, Oxford.

Getz, D. and Page, S.J. (2016), "Progress and prospects for event tourism research", *Journal of Tourism Management*, Vol. 52, pp. 593–631.

Heeley, J. (2011), *Inside City Tourism: A European Perspective*. Channel View, Bristol.

Higgins, R. (2017), "The rise of nontraditional event venues" [online] https://www. eventbrite.com/blog/the-rise-of-nontraditional-event-venues-ds00/

Hobbs, T. (2015), "Instagram to become 'shop window for brands' this Christmas as it debuts new ad targeting tool" [online] https://www.marketingweek.com/ 2015/11/25/instagram-to-become-shop-window-for-brands-this-christmas-as-it-launches-new-ad-targeting-tool/

Hobswarm, E. and Ranger, T. (1983), *The Invention of Tradition*. Cambridge University Press.

Howard, E. (2007), "New shopping centres: is leisure the answer?", *International Journal of Retail & Distribution Management*, Vol. 35, No. 8, pp. 661–672.

Jago, L., Chalip, L., Brown, G., Mules, T. and Ali, S. (2003), "Building events into destination branding", *Journal of Event Management*, Vol. 8, pp. 3–14.

Jones, M.A. (1999), "Entertaining shopping experiences: an exploratory investigation", *Journal of Retailing and Consumer Services*, Vol. 6, No. 3, pp. 129–139.

Jones, M.A., Reynolds, K. and Arnold, M.J. (2006), "Hedonic and utilitarian shopping value: investigating differential effects on retail outcomes", *Journal of Business Research*, Vol. 59, No. 9, pp. 974–981.

Leask, A. (2010), "Progress in visitor attraction research: towards more effective management", *Tourism Management*, Vol. 31, pp. 155–166.

Litvin, S.W. (2007), "Marketing visitor attractions: a segmentation study", *International Journal of Tourism Research*, Vol. 9, pp. 9–19.

Lombart, C. (2004), "Browsing behaviour in retail stores: an opportunity for retailers?" [online] https://www.highbeam.com/doc/1G1-128606453.html (accessed 2 September 2017).

Mall of Qatar (2017), "Grand opening party" [online] http://www.mallofqatar.com. qa/english/events/mall-of-qatar-grand-opening

Mall of the Emirates (2017), "Ski Dubai" [online] http://www.malloftheemirates. com/entertainment/ski-dubai

Meadowhall (2017), "What's on" [online] https://www.meadowhall.co.uk/Artisan

Mehta, S., Jain, A. and Jawale, R. (2014), "Impact of tourism on retail shopping in Dubai", *International Journal of Trade, Economics and Finance*, Vol. 5, No. 6, pp. 530–535.

Michelli, J. (2007), *The Starbucks Experience*. McGraw-Hill, New York.

Molloy, M. (2016), "'World's saddest polar bear' kept in shopping centre finally sees some sunlight", *The Telegraph* [online] http://www.telegraph.co.uk/news/2016/11/ 14/worlds-saddest-polar-bear-kept-in-shopping-centre-finally-sees-s/

Morgan, N., Pritchard, A. and Pride, R. (2011), *Destination Brands: Managing Place Reputation*, 3rd ed. Butterworth-Heinemann, Oxford.

O'Toole, W. (2011), *Events Feasibility and Development: From Strategy to Operations*. Elsevier Butterworth-Heinemann, Oxford.

Picsolve (2015), "Insight Report: UAE's Attraction and Theme Park Industry" [online] http://www.picsolve.biz/files/2514/3168/6982/Picsolve_Insights_Report_UAE_2015. pdf

Pike, S. (2012). *Destination Marketing*. Routledge, Abingdon.

Pine, J.P. II and Gilmore, J.H. (1999), "Welcome to the experience economy", *Harvard Business Review*, Vol. 76, No. 4, pp. 97–106.

Richards, G.W. and Palmer, R. (Eds.). (2010), "Eventful cities", *Cultural Management and Urban Revitalisation*, Butterworth-Heinemann, Oxford.

Sharples, L., Crowther, P., May, D. and Orefice, C. (2014), *Strategic Event Creation*. Goodfellow Publishers Ltd, Oxford.

Swarbrooke, J. (2015), *The Development and Management of Visitor Attractions*. Routledge, London.

Thomassen, L., Lincoln, K. and Aconis, A. (2006), *Retailization: Brand Survival in the Age of Retailer Power*. Kogan Page, Philadelphia.

Tressider, R. and Hirst, C. (2012), *Marketing in Food, Hospitality, Tourism and Events: A Critical Approach*. Goodfellow Publishers Ltd, Oxford.

Tribe, J. (2011), *The Economics of Recreation, Leisure and Tourism*. Butterworth-Heinemann, Oxford.

Underhill, P. (2000), *Why We Buy: The Science of Shopping*. Touchstone, New York.

Urry, J. (1995). *Consuming Places*. Routledge, London.

Wakefield, K.L. and Baker, J. (1998), "Excitement at the mall: determinants and effects on shopping response", *Journal of Retailing*, Vol. 74, No. 4, pp. 515–539.

Wang, Y. and Pizam, A. (2011), *Destination Marketing and Management: Theories and Applications*. CABI, Wallingford.

Whitfield, J. (2009), "Why and how UK visitor attractions diversify their product to offer conference and event facilities", *Journal of Convention & Event Tourism*, Vol. 10, No. 1, pp. 72–88.

Woodruffe-Burton, H., Eccles, S. and Elliot, R. (2001), "Towards a theory of shopping: a holistic framework", *Journal of Consumer Behaviour*, Vol. 1, No. 3, pp. 256–266.

Ziakas, V. (2013), "A multi-dimensional investigation of a regional event portfolio: advancing theory and praxis", *Journal of Event Management*, Vol. 17, No. 1, pp. 27–48.

Ziakas, V. (2014), "Planning and leveraging event portfolios: towards a holistic theory", *Journal of Hospitality Marketing and Management*, Vol. 23, No. 3, pp. 327–356.

3 Managing the event workforce

Analysing the heterogeneity of job experiences

Birgit Muskat and Judith Mair

Introduction

This chapter contributes to a deeper understanding of the event workforce. The heterogeneity of job experiences is analysed; whereas most event research focuses on understanding the event activity, event visitors and event impacts, an 'inward' perspective is adopted. Thus, the organisational context of event workforce is centrestage as this 'inward' perspective remains largely ignored in the events research domain (see for example Liu, 2018; Mair & Whitford, 2013; Muskat & Deery, 2017). Specifically, the focus is on analysing the underlying intrinsic and extrinsic motivators in the heterogeneous event workforce. A critical examination of job satisfaction is highly important as satisfied employees are healthier (Bowling et al., 2010), more positive towards their organisation, more effective (Judge et al., 2017) and committed (Tsai & Yen, 2018), and more easily develop positive relationships with their customers (Menguc et al., 2016).

Job satisfaction is the result of an individual's appraisal of various factors of overall job experience (Jeanson & Michinow, 2018; Locke, 1969). Moreover, a contextual understanding of job experiences and satisfaction in the events context is as important as in the human resources and organisational behaviour domain, both of which has shown how job satisfaction positively influences employee's attitudes, commitment, well-being (Bowling et al., 2010; Tsai & Yen, 2018; Wright & Cropanzano, 2000) and results in better team and leadership performance, as well as organisational effectiveness (Braun et al., 2013; Judge et al., 2017).

Traditionally, the event research domain has been strongly 'outward-looking', e.g. event visitor numbers and economic impacts are in the centre of interest. However, literature in the events area is gradually taking a more critical turn, with a rejection of the instrumentalist and neoliberal discourses around event 'management' and a move towards a better, and more holistic, understanding of how events can and should be understood (Lamond & Platt, 2016). 'Inward' perspectives – organisational and staff perspectives – to date have been rather ignored, which might be routed in the overall lack of interest in understanding the enabler side in the tourism

(Zehrer et al., 2014) or hospitality sector (Solnet et al., 2016). In addition, we argue that existing literature has failed to adequately take into account the significant diversity in event workforce.

To fill this gap, we posit to understand the factors that make the role of event managers distinctive or provide insights for employees and volunteers on how to perform best in this unique workplace setting. We address the overlooked research question: *What constitutes job satisfaction for the event workforce?* Job satisfaction is the result of an individual's appraisal of various factors of an overall job experience (Jeanson & Michinow, 2018; Locke, 1969; Sirgy et al., 2001). We argue that job satisfaction for event staff requires a deeper understanding of the unique characteristics of events including the pulsating nature, the temporary and fast-paced dimension and the exceptional heterogeneity of the workforce that shapes organisational processes. The unique organisational context that is created shapes the individual event-job experiences and consequently determines job satisfaction for both event employees and volunteers.

Subsequently, our aim is to provide insights that are relevant to event managers, specifically in relation to job experiences and job satisfaction management. We synthesise the unique organisational aspects of events and theoretically explore how the event workforce, consisting of permanent employees, casual employees and unpaid volunteers, experiences job satisfaction. In doing so, we contribute towards the move to a less instrumentalist and more holistic understanding of events and the employees and volunteers that run them. In this chapter, we focus mainly on major events which attract tourists to a destination, and which are large enough to have a complement of full-time permanent employees. However, elements of the arguments in this chapter will also have relevance to smaller events, even those entirely organised and staffed by volunteers. Given the growing number of events and the resulting need to professionally manage the event workplace, implications for management practices are discussed.

Job satisfaction and job experiences at events

Job satisfaction is formed through a number of components that matter to each employee and is created through the appraisal of attributes of the individual's job experience (Jeanson & Michinow, 2018; Locke, 1969; Sirgy et al., 2001). Motivation theory can be used to explicate underlying needs and values that motivate and influence job experience (Deci & Ryan, 1985; Tietjen & Myers, 1998). The identification of a staff member's motives helps us to determine what drives them to pursue certain goals – job satisfaction can be attained by achieving these goals (Gagné & Deci, 2005). Sirgy et al. (2001) explain that when employees' values and needs are met, the behavioural response of 'relaxation' (job satisfaction) occurs; however, non-fulfilment of needs leads instead to stress and tensions – and ultimately job dissatisfaction.

To better understand these motivational factors, Porter and Lawler's (1968) dichotomy of intrinsically and extrinsically controlled factors of motivation has often been used to categorise concrete factors that accumulate towards job satisfaction (Borzaga & Tortia, 2006; Kuvaas et al., 2017; Lee et al., 2015). *Intrinsic motivational factors* are based on a staff member's inner needs, values and beliefs. Amabile et al. (1994) explain that intrinsic motivational factors include being self-determinant at work, working autonomously, mastery and challenges related to work, involvement in tasks, as well as curiosity, enjoyment related to work and the work environment. In contrast, a staff member's *extrinsic motivational factors* are externally controlled and related to the evaluation by the organisation. Factors include recognition from the organisation, competition, money or other tangible incentives (Amabile et al., 1994). Whereas intrinsic motivation is difficult for an organisation to influence, extrinsic motivation is largely controlled by organisations. Extrinsic reward systems can, for example, include monetary bonuses, career advancement and official recognition. Recognition that leads to career progress (Scarpello, & Campbell, 1983) can also be an important factor for job satisfaction, too – especially if career progress is of importance to the staff member.

Organisational behaviour literature confirms that high levels of staff job satisfaction are significant for both the individual and the organisation. Satisfied employees are healthier, more satisfied with their lives and show higher levels of well-being (Bowling et al., 2010; Wright & Cropanzano, 2000). When organisations meet or even exceed their workforce's expectations, staff are satisfied and subsequently develop a generally positive attitude towards their organisation. This, in turn, leads to positive behavioural outcomes such as higher organisational effectiveness (Judge et al., 2001, 2017; Koys, 2001), higher commitment (Tsai & Yen, 2018), better team and leadership performance (Braun et al., 2013), and improved citizenship behaviour (Koys, 2001). Importantly, the marketing literature also confirms that there is a positive relationship between satisfied employees and satisfied customers (Menguc et al., 2016), which is particularly important for events, with close customers and event staff interactions.

Studies suggest that there are differences in job satisfaction between employees and volunteers when doing similar tasks (Pearce, 1983). Authors also find that achieving high levels of job satisfaction for volunteers is even more important for volunteers than for employees, when predicting organisational commitment and increases their intention to stay at the event (Bang, 2015; Bang et al., 2012; Williams & Anderson, 1991) and fosters volunteer's engagement at the event (Millette & Gagné, 2008). For the majority of volunteers, the key antecedent to job satisfaction is autonomy; yet Bang (2015) notes that there is a difference between young and older volunteers, and job satisfaction is more important to retain older volunteers than younger volunteers.

The event organisation: the pulsating nature

The chapter now turns to discuss the unique organisational context of events that is likely to influence the underlying motivational factors leading to job satisfaction. As mentioned in the introduction, the focus of this chapter is on major events, defined as "events that are capable, by their size and media interest, of attracting significant visitor numbers, media coverage and economic benefits" (Allen et al., 2011, p. 14). The first important and unique organisational characteristic is the *pulsating nature* of the events organisation (Aisbett, & Hoye, 2014; Deery & Jago, 2005; Hanlon & Jago, 2004). The pulsating nature is a result of the different phases of event management (Muskat & Deery, 2017). The pre-event phase is shaped by planning activities, training and knowledge sharing, and recruiting of the event workforce. The event operations phase is characterised by the instant need of the workforce to collaborate with each other, regardless of whether they have known each other or collaborated before or not. High speed, quick and effective decision making is needed during this relatively short phase to perform successfully. After the event and with the commencement of the post-event phase, most casual and volunteer workforces exit the event workforce immediately after the event.

Consequently, after the departure of most of the event workforce, only the permanent workforce remains and often there is limited time to reflect and celebrate and recognise achievements in this phase, as the planning for the next upcoming event has just started (Muskat & Deery, 2017). Most events operate with some full-time permanent employees but a large number of casual and volunteers join the team during the event operations phase. While this is the case to a certain extent for most events, this pulsating nature is particularly relevant for major events that have a strong relationship with tourism in a destination. Hence, a small skeleton staff work for the event for majority of the time, a substantial swelling to large numbers of staff with various contract forms occurs in the run-up to and during the event, and finally a dropping away again to very small numbers takes place once the event is over (Hanlon & Jago, 2004). Consequently, this pulsating nature determines staff fluctuations and influences social aspects, including team cohesion, recognition of joint achievements and organisational learning.

The event activity: *temporary, fast-paced and short-term*

The second unique characteristic of the events workplace is the *temporary, fast-paced* and *short-term dimension* of the event activity. Yeoman et al. (2012) point out that the event activity, the organisational processes involved, and features of the workforce are both temporary and ever changing. Whilst this can potentially be a positive characteristic, given that temporary organisations tend to have clear timelines and are well organised (Parent and McIntosh, 2013), there are resulting difficulties that need to be reflected upon and managed by event staff. Another implication of the temporary dimension arises in that event teams need to be put together at fairly short notice and work intensively together, yet only for a limited period of time (Parent & McIntosh, 2013); thus quick problem solving is a key task (Getz & Page, 2016).

The combination of the short-term, temporary organisational structure and the demands of the work are reflected in a short-term orientation among team members. Moreover, and presumably as a consequence of the pace of the event activity, the event workforce is recognised as having to work extremely hard under stressful conditions (e.g. Van der Wagen & White, 2014). Arguably, these job characteristics make it even more difficult for event managers to find the time to reflect on the motivation of their staff, which is an important factor in levels of job satisfaction within diverse teams. Event managers also have little time to find effective measures that increase job satisfaction, taking into account the different values and attitudes of various team members.

In summary, both the pulsating nature of events, on the one hand, and the temporary and short-term dimension of the event, on the other hand, mean that it can be very difficult for event organisations to attract and retain high performing permanent and casual staff, as well as volunteers for this face-paced work environment. Attracting and above all retaining suitable and experienced permanent, casual employees and volunteers – and meeting their job expectations and understanding their motivation – are presumably challenging. The time pressures and demands on event organisers mean an added layer of complexity in terms of management. There are some overlaps between events and other types of tourism business, for example in relation to seasonal tourism businesses, which also have to deal with fluctuating staff numbers; however, such fluctuations are significantly more dramatic in the events context.

Event teams: the heterogeneity

This section discusses *heterogeneity of event teams*, the third factor that makes the event organisational context unique and needs to be considered in terms of the event workforce's job satisfaction. We argue that with heterogeneity comes a variety of different values, beliefs and needs of employees and volunteers that make motivational factors and subsequently job experiences and job satisfaction complex to manage. In order to analyse what constitutes heterogeneity in event teams, we first review the organisational behaviour literature on heterogeneity in teams and second, we analyse heterogeneity as it relates to events. Heterogeneous teams are often desired in workforce composition; they are more creative and often outperform homogeneous teams in terms of performance (Jackson, 1996; Katzenbach & Smith, 2015). Yet, many organisations perceive that diversity runs in tandem with unproductive processes and can also lead to negative, ineffective outcomes

Table 3.1 Heterogeneity in work teams

Dimension of diversity	Examples	References
Demographic heterogeneity	Diverse age, citizenship, ethnicity, gender	Chatman & Flynn (2001)
Functional heterogeneity	Diverse specialisations and work roles in the team	Jackson (1996); Bunderson & Sutcliffe (2002)
Task-related heterogeneity	Different educational level, formal credentials, knowledge and expertise, task experience	Jackson et al. (1995); Jackson (1996)
Relations-oriented heterogeneity	Diverse attitudes, values, personality	Jackson et al. (1995); Jackson (1996)
Career-related heterogeneity	Similar experiences or different past experiences, in jobs, work environments, affiliations, organisations	Beckman et al. (2007)

(Nielsen Tasheva & Hillman, 2018). Problems in functionally heterogeneous teams are often a lower willingness to share information, a higher level of dysfunctional conflict and high staff turnover rates, which are all the result of higher negative job experiences and lower job satisfaction (Bunderson & Sutcliffe, 2002). As depicted in Table 3.1, the prevailing literature in management suggests there are five dimensions of heterogeneity in work teams: demographic-, functional-, task-related dimensions of heterogeneity – which might be considered to be more explicit and easier to capture for both employees and managers – and further career- and relations-oriented heterogeneity, which might rather be subtler and less visible:

> What managers (and some researchers) often ignore are the possible effects of the relations-oriented diversity that might be present in such a team. Relations-oriented diversity can shape behavior even when there is no association between it and the team's task-related attributes, because it triggers stereotypes that influence the way team members think and feel-about themselves as well as others on the team.
>
> (Jackson, 1996, p. 57)

In relation to events, it becomes clear that the event workforce consists of high levels of heterogeneity; functional heterogeneity in particular might be higher when compared to other, more traditional and long-term oriented organisations. Functional heterogeneity also arises due to different contract forms of the event workforce. Mair (2009) notes that the event workforce is often made up of various alternative forms of employees. These include full-time and part-time work for long-term and/or permanent paid employees, short-term casual employees, recurring volunteers and first-time and long-time volunteers, as well as staff recruited and managed by a range of external contractors (such as security and catering staff amongst others). Event staff might have full-time or part-time contracts, might be casual/temporary paid staff or might be volunteers; all of them have different levels of expertise, motivation and expectations about their event job experience – so by extension, they will have different job satisfaction. In many businesses, managers have to deal with a variety of staff types. For example, tourist enterprises such as Visitor Information Centres may have a mix of volunteer and paid employees. However, the full gamut of heterogeneity in the events context is almost unique.

This functional heterogeneity clearly indicates the likelihood of differences in job motivation and drivers of job satisfaction between permanent and casual employees or volunteers (Bernhard-Oettel et al., 2005; Muskat & Deery, 2017). Full-time and *permanent employees* are often in managerial roles and provide continuity in the long run for the event organisation. For permanent employees, building long-term careers is likely to be one of the key motives to join the event. Motivations of the permanent event workforce might be related to building an event career due to the fast-paced and vibrant job environment but also in long-term capacity and competence building, as well as earning rewards and recognition from the event organisation. This is arguably more likely to be the case for major tourist-attracting events that recur each year, such as the Edinburgh Festival, the Munich Oktoberfest or the Rio Carnival, as they are more likely to have employment openings for full-time permanent employees, and career progression opportunities. Smaller events, such as those organised primarily for local communities, are more likely to be fully organised and run by volunteers, thus not offering such employment or career opportunities. Events can be organised in both for-profit and non-profit context. Yet, for non-profit workplaces, research has found that stakeholders tend to have a higher degree of intrinsic motivation. This is because, traditionally, compensation levels are below market average and non-profit employees are more likely to report that their work is more important to them than the money they earn (Mirvis & Hackett, 1983).

In contrast, *casual employees* are more likely to have short-term interests or expectations about their work for the event. Casual employees at events are hired as the need arises due to high demand in personnel during the event operations phase. Casual employment is less stable and often associated with less favourable job characteristics (OECD, 2002). The OECD further notes that casual employees usually receive less pay and benefits, and generally receive neither paid vacations nor unemployment insurance (OECD, 2018). Yet, there might also be a number of positive aspects in being employed on a casual basis such as more mobility and higher flexibility for the individual to explore and enter new working areas. Casual employees might also be interested in pursuing a long-term event career and opt to screen a potential employer for suitability, before then deciding whether or not to apply for a permanent job.

Thus, the motivation for casual employees to join an event might be to benefit their personal preferences, e.g. they might prefer flexibility in terms of daily work but may also be related to the terms and flexibility of their working contract. Specifically, research has found that flexibility in working hours and having the opportunity to work extensively – and after hours outside the standard weekly hours – are often key motivators for casual employees (Gottschalk & McEachern, 2010). Other motives for casual personnel to join events might be the opportunity for entry or re-entry into the workforce (Pocock et al., 2009). However, it is unlikely that casual employees are motivated by or committed to the event organisation itself, as their commitment is short-term driven. This is, of course, unless the event organisation provides casual employees with an opportunity to build pathways into more long-term employment.

The third group that makes up the event workforce are *unpaid volunteers.* Volunteers have been referred to as "the life blood" of events (Goldblatt, 2002, p. 110) and often highly outnumber paid event personnel, especially during the operations phase of the event. Research has shown that volunteers have different motives to join the event workforce (Holmes et al., 2010). Volunteers for example are motivated by the possibility to gain skills and competencies that lead to career advantages, and they also use event experiences as a transitioning period from unemployment to employment – similar to casual employees (Pocock et al., 2009). In some cases, particularly for smaller community events, the entire event organisational team consists of volunteers, thus providing a wider range of skills development for these volunteers (Holmes et al., 2010). Yet, other volunteers might simply seek a purely leisure and recreational experience from joining the event team (Mojza et al., 2010).

Importantly, volunteers are not a homogeneous group; instead, motivations to join the event differ within the volunteer group (Clary et al., 1998). For example, Muskat and Deery (2017) found that different motivations lead to different workplace behaviours and different willingness to share knowledge and collaborate during events. Whereas inexperienced volunteers wanted to gain new knowledge and learn, more senior and experienced volunteers were driven by self-actualisation and would rather experiment or do things in their preferred way. Hence, the more experienced volunteers did not want to adhere to rules and guidelines and had less interest in collaboration and knowledge sharing. Younger volunteers instead were motivated by gaining competencies that would serve them for their future careers (Muskat & Deery, 2017).

The literature outside of the event volunteering domain offers a number of motivations for volunteering. For example, one of the seminal works on volunteer motivations was the Voluntary Functions Inventory (VFI) presented by Clary et al. (1998). This took a functional approach, to try to examine what function volunteering played for those who give up their time. They concluded that there were six dimensions: (i) *values* (volunteering allows an individual to express values related to altruistic and humanitarian concerns for other), (ii) *understanding* (volunteering permits new learning experiences and the chance to expand knowledge, skills and abilities), (iii) *social* (volunteering gives one opportunities to be with one's friends), (iv) *career* (volunteering can have career-related benefits), (v) *protective* (volunteering may reduce guilt over being more fortunate than others) (vi) and *enhancement* (volunteering can allow someone to feel good about themselves). Other works (see for example Edwards, 2005) identified different dimensions in the museum volunteering context. These dimensions included personal needs, relationship network, self-expression, available time, social needs, purposive needs, free time and personal interest (in the museum, or in one aspect of the collection).

Conclusion

In this chapter we have discussed an under researched area in the events' management research domain and concluded that whilst event researchers have shown a great interest in understanding event consumers, the enabler side – employee – has been neglected. In summary, we posit that the pulsating nature of event organisation, the temporary, fast-paced and short-term dimension of the event activity, and the heterogeneous workforce form a unique organisational context for events that have implications for event management. Understanding these characteristics is important, as studies have shown that they affect the individual and lead to different workplace behaviour outcomes during and after events (Muskat & Deery, 2017). In addition, the chapter has shown the significant diversity that exists within the event workforce, thus bringing a more critical and holistic approach to event management research. The arguments in this chapter are more broadly relevant to larger scale major events that play an important tourist-attracting role in a destination, although elements of our argument are likely to be equally relevant for the organisation and management of smaller community events. We now respond to the set research question: *what constitutes job satisfaction for the event workforce?* We have also given a summary of how organisational context influences job satisfaction for the event workforce.

First, we suggest, for the event organisational context, job satisfaction for *permanent employees* might be a combination of both intrinsic, e.g. mastery of the challenge of the fast-paced work environment, and extrinsic motivational factors, e.g. recognition, incentive of attending the event, career progress. Hence, we posit that the fast-paced nature of the event work environment, the pace and the vibrant atmosphere constitute additional key aspects that are relevant for job satisfaction. However, whilst these motivating factors might be relevant to those joining the event workforce in the first place, we argue that over time they might also become barriers to job satisfaction. High stress levels due to constant exposure to a rapid and dynamic workplace and changing tasks and workforce might affect job satisfaction in the long term. Practically, event managers could enhance job satisfaction particularly during the post-event phase – appreciation of work with rewards and recognition, and by seeking and responding to feedback (Sirgy et al., 2001). Further, event managers should be aware that high stress levels and an ever-changing workforce might require some extra activities around personal resilience and mindfulness of maintaining staff work–life balance.

Second, for *casual event employees*, job satisfaction is likely to be different again. Casual employees are likely to have rather short-term interests. Hence, key motivators might be flexibility in working hours and even having the opportunity to work extensively for some time, to free up time for other periods. Drawing on the literature, we posit that earning money and time flexibility are major extrinsic motivators for casual employees, whereas curiosity in relation to the task and work environment may serve as important intrinsic motivators (Amabile et al., 1994). However, some casual employees might join for similar motives as some volunteers. For example, for some casual employees, the event experience might be a pathway to permanent work, and learning opportunities might be important factors for job. Most importantly, however, there is no long-term commitment that links casual employees to the event; hence, to retain high performing staff, event managers need to make an effort in managing their job satisfaction, as ties are certainly weaker for this group of the workforce.

Third, *unpaid volunteers* have even more complex underlying drivers that lead to their job satisfaction. It is clear that job satisfaction between employees and volunteers when doing similar tasks (Pearce, 1983) is different. Volunteers, who are satisfied with their jobs, show higher commitment, engage more in their task and are even more likely to return and volunteer for the next event (e.g. Bang, 2015; Millette & Gagné, 2008). Most evidently, the key difference between volunteers and employees is that money is not a motivating factor. Yet for volunteer's high levels of job satisfaction, other extrinsic factors might be even more important, including recognition and potential career advancement. Intrinsic motivators might be manifold and might include autonomous work, mastery and the appreciation of challenge, as well as curiosity of the work environment and enjoyment. In fact, the need to engage in autonomous work is one of the key predictors for volunteer's job satisfaction (Boezeman & Ellemers, 2009). For event managers, understanding these predictors is essential, as events activities cannot be successful without volunteers (Goldblatt, 2002), and volunteers usually highly outnumber the paid permanent and casual event workforce, and job satisfaction for volunteers has been found to be an essential factor to predict the intention to continue engaging in events. It is notable though that event volunteers are generally interested in the work and the experience. However, organisational commitment is likely to be less strong for event volunteers when compared with permanent event employees. In order to retain event volunteers, managers should consequently understand the drivers of volunteer job satisfaction and the importance of volunteer relationship management.

Overall, we conclude that so far event researchers have rather neglected to provide critical insights and a deeper understanding of employees and volunteers. Our study posits that demographic, functional, task-related, relation-oriented and career-related heterogeneity of event staff shapes underlying and diverse motivations; they presumably have distinct and perhaps competing values and attitudes, which in consequence influence job satisfaction. Departing from this, we suggest that future research should conduct studies that provide empirical data to better understand this rather neglected heterogeneous nature of the event workforce. Specifically, future research studies should use motivation theory to understand differences in motives to join events, between permanent, casual employees and volunteers. Further, studies should evaluate which leadership styles influence the event workforce best or explore what constitutes leadership in this fast-paced, heterogeneous work environment.

In terms of practical implications, we recommend that event managers need to actively manage job satisfaction and the job experiences of their heterogeneous workforce. Establishing routines and mechanisms, for example, might enable to understand, reflect and communicate job experiences and manage job satisfaction. Both the understanding and creation of practices are essential, as job satisfaction leads to improved organiser effectiveness (Judge et al., 2017), higher commitment of the workforce (Tsai & Yen, 2018), retention of both paid employees and volunteers (Bang, 2015; Millette & Gagné, 2008) – and ultimately to more satisfied event visitors (Menguc et al., 2016).

References

Aisbett, L., & Hoye, R. (2014). The nature of perceived organizational support for sport event volunteers. *Event Management, 18*(3), 337–356.

Allen. J., O'Toole, W., Harris, R., & McDonnell, I. (2011). *Festival and Special Event Management*. Milton, QLD: John Wiley & Sons.

Amabile, T. M., Hill, K. G., Hennessey, B. A., & Tighe, E. M. (1994). The work preference inventory: Assessing intrinsic and extrinsic motivational orientations. *Journal of Personality and Social Psychology, 66*(5), 950–967.

Bang, H. (2015). Volunteer age, job satisfaction, and intention to stay: A case of nonprofit sport organizations. *Leadership & Organization Development Journal, 36*(2), 161–176.

Bang, H., Ross, S., & Reio Jr, T. G. (2012). From motivation to organizational commitment of volunteers in non-profit sport organizations: The role of job satisfaction. *Journal of Management Development, 32*(1), 96–112.

Beckman, C. M., Burton, M. D., & O'Reilly, C. (2007). Early teams: The impact of team demography on VC financing and going public. *Journal of Business Venturing, 22*(2), 147–173.

Bernhard-Oettel, C., Sverke, M., & De Witte, H. (2005). Comparing three alternative types of employment with permanent full-time work: How do employment contract and perceived job conditions relate to health complaints? *Work & Stress, 19*(4), 301–318.

Boezeman, E. J., & Ellemers, N. (2009). Intrinsic need satisfaction and the job attitudes of volunteers versus employees working in a charitable volunteer organization. *Journal of Occupational and Organizational Psychology, 82*(4), 897–914.

Borzaga, C., & Tortia, E. (2006). Worker motivations, job satisfaction, and loyalty in public and nonprofit social services. *Nonprofit and Voluntary Sector Quarterly, 35*(2), 225–248.

Bowling, N. A., Eschleman, K. J., & Wang, Q. (2010). A meta-analytic examination of the relationship between job satisfaction and subjective well-being. *Journal of Occupational and Organizational Psychology, 83*(4), 915–934.

Braun, S., Peus, C., Weisweiler, S., & Frey, D. (2013). Transformational leadership, job satisfaction, and team performance: A multilevel mediation model of trust. *The Leadership Quarterly, 24*(1), 270–283.

Bunderson, J. S., & Sutcliffe, K. M. (2002). Comparing alternative conceptualizations of functional diversity in management teams: Process and performance effects. *Academy of Management Journal, 45*(5), 875–893.

Chatman, J. A., & Flynn, F. J. (2001). The influence of demographic heterogeneity on the emergence and consequences of cooperative norms in work teams. *Academy of Management Journal, 44*(5), 956–974.

Clary, E., Ridge, R., Stukas, A., Snyder, M., Copeland, J., Haugen, J., & Miene, P. (1998). Understanding and assessing the motivations of volunteers: A functional approach. *Journal of Personality and Social Psychology, 74*(96), 1516–1530.

Deci, E. L., & Ryan, R. M. (1985). The general causality orientations scale: Self-determination in personality. *Journal of Research in Personality, 19*(2), 109–134.

Deery, M., & Jago, L. (2005). The management of sport tourism. *Sport in Society, 8*(2), 378–389.

Edwards, D. (2005). It's mostly about me: Reasons why volunteers contribute their time to museums and art museums. *Tourism Review International, 9*(1), 21–31.

Gagné, M., & Deci, E. L. (2005). Self-determination theory and work motivation. *Journal of Organizational Behavior, 26*(4), 331–362.

Getz, D., & Page, S. (2016). *Event Studies: Theory, Research and Policy for Planned Events* (3rd ed.). London: Routledge.

Goldblatt, J. (2002). *Special Events: 21st Century Global Events Management* (3rd ed.). New York, NY: Wiley.

Gottschalk, L., & McEachern, S. (2010). The frustrated career: Casual employment in higher education. *Australian Universities' Review, 52*(1), 37–50.

Hanlon, C., & Jago, L. (2004). The challenge of retaining personnel in major sport event organizations. *Event Management, 9*(1–2), 39–49.

Holmes, K., Smith, K. A., Lockstone-Binney, L., & Baum, T. (2010). Developing the dimensions of tourism volunteering. *Leisure Sciences, 32*(3), 255–269.

Jackson, S. E. (1996). The consequences of diversity in multidisciplinary work teams. In M. A. West (Ed.), *Handbook of Work Group Psychology* (pp. 53–75). Chichester: Wiley.

Jackson, S. E., May, K. E., & Whitney, K. (1995). Understanding the dynamics of diversity in decision-making teams. In R. A. Guzzo, E. Salas. & Associates (Eds.), *Team Effectiveness and Decision Making in Organizations* (pp. 204–226). San Francisco, CA: Jossey Bass Publishers.

Jeanson, S., & Michinov, E. (2018). What is the key to researchers' job satisfaction? One response is professional identification mediated by work engagement. *Current Psychology* (online-before-print), 1–10.

Judge, T. A., Thoresen, C. J., Bono, J. E., & Patton, G. K. (2001). The job satisfaction–job performance relationship: A qualitative and quantitative review. *Psychological Bulletin, 127*(3), 376–407.

Judge, T. A., Weiss, H. M., Kammeyer-Mueller, J. D., & Hulin, C. L. (2017). Job attitudes, job satisfaction, and job affect: A century of continuity and of change. *Journal of Applied Psychology, 102*(3), 356–374.

Katzenbach, J. R., & Smith, D. K. (2015). *The Wisdom of Teams: Creating the High-Performance Organization.* Boston, MA: Harvard Business Review.

Koys, D. J. (2001). The effects of employee satisfaction, organizational citizenship behavior, and turnover on organizational effectiveness: A unit-level, longitudinal study. *Personnel Psychology, 54*(1), 101–114.

Kuvaas, B., Buch, R., Weibel, A., Dysvik, A., & Nerstad, C. G. (2017). Do intrinsic and extrinsic motivation relate differently to employee outcomes? *Journal of Economic Psychology, 61*, 244–258.

Lamond, I. R., & Platt, L. (2016). Introduction. In I. R. Lamond & L. Platt (Eds.), *Critical Event Studies: Approaches to Research.* London: Macmillan.

Lee, J. S., Back, K. J., & Chan, E. S. (2015). Quality of work life and job satisfaction among frontline hotel employees: A self-determination and need satisfaction theory approach. *International Journal of Contemporary Hospitality Management, 27*(5), 768–789.

Liu, C. H. S. (2018). Examining social capital, organizational learning and knowledge transfer in cultural and creative industries of practice. *Tourism Management, 64*, 258–270.

Locke, E. A. (1969). What is job satisfaction? *Organizational Behavior and Human Performance, 4*(4), 309–336.

Mair, J. (2009). The events industry: The employment context. In T. Baum, M. Deery, C. Hanlon, L. Lockstone, & K. Smith (Eds.), *People and Work in Events and Conventions: A Research Perspective.* Wallingford, OX: CABI.

Mair, J., & Whitford, M. (2013). Special issue on event and festival research methods and trends. *International Journal of Event and Festival Management, 4*(1), 6–30.

Menguc, B., Auh, S., Katsikeas, C. S., & Jung, Y. S. (2016). When does (mis)fit in customer orientation matter for frontline employees' job satisfaction and performance? *Journal of Marketing, 80*(1), 65–83.

Millette, V., & Gagné, M. (2008). Designing volunteers' tasks to maximize motivation, satisfaction and performance: The impact of job characteristics on volunteer engagement. *Motivation and Emotion, 32*(1), 11–22.

Mirvis, P. H., & Hackett, E. J. (1983). Work and work force characteristics in the nonprofit sector. *Monthly Labor Review, 106*(4), 3–12.

Mojza, E. J., Lorenz, C., Sonnentag, S., & Binnewies, C. (2010). Daily recovery experiences: The role of volunteer work during leisure time. *Journal of Occupational Health Psychology, 15*(1), 60–74.

Muskat, B., & Deery, M. (2017). Knowledge transfer and organizational memory: An events perspective. *Event Management, 21*(4), 431–447.

Nielsen Tasheva, S., & Hillman, A. (2018). Integrating diversity at different levels: Multi-level human capital, social capital, and demographic diversity and their implications for team effectiveness. *Academy of Management Review* (online-before-print), *444* (4) 746–765.

OECD (2002). *Taking the Measure of Temporary Employment.* Retrieved from https://www.oecd.org/employment/emp/17652675.pdf

OECD (2018). *Temporary Employment (Indicator)*. Retrieved from https://data. oecd.org/emp/temporary-employment.htm

Parent, M. M., & MacIntosh, E. W. (2013). Organizational culture evolution in temporary organizations: The case of the 2010 Olympic Winter Games. *Canadian Journal of Administrative Sciences / Revue Canadienne Des Sciences De l'Administration, 30*(4), 223–237.

Pearce, J. L. (1983). Job attitude and motivation differences between volunteers and employees from comparable organizations. *Journal of Applied Psychology, 68*(4), 646–652.

Pocock, B., Skinner, N., & Ichii, R. (2009). *Work, Life and Workplace Flexibility (AWAU): The Australian Work and Life Index*. Retrieved from University of South Australia website: unisa.edu.au/hawkeinstitute/cwl

Porter, L. W., & Lawler, E. E. (1968). What job attitudes tell about motivation. *Harvard Business Review, 46*(1), 118–126.

Scarpello, V., & Campbell, J. P. (1983). Job satisfaction and the fit between individual needs and organizational rewards. *Journal of Occupational and Organizational Psychology, 56*(4), 315–328.

Sirgy, M. J., Efraty, D., Siegel, P., & Lee, D. J. (2001). A new measure of quality of work life (QWL) based on need satisfaction and spillover theories. *Social Indicators Research, 55*(3), 241–302.

Solnet, D., Baum, T., Robinson, R. N., & Lockstone-Binney, L. (2016). What about the workers? Roles and skills for employees in hotels of the future. *Journal of Vacation Marketing, 22*(3), 212–226.Tietjen, M. A., & Myers, R. M. (1998). Motivation and job satisfaction. *Management Decision, 36*(4), 226–231.

Tsai, C. F., & Yen, Y. F. (2018). Moderating effect of employee perception of responsible downsizing on job satisfaction and innovation commitment. *The International Journal of Human Resource Management* (online-before-print), 1–25.

Van der Wagen, L., & White, L. (2014). *Human Resource Management for the Event Industry* (2nd ed.). London: Routledge.

Williams, L. J., & Anderson, S. E. (1991). Job satisfaction and organizational commitment as predictors of organizational citizenship and in-role behaviors. *Journal of Management, 17*(3), 601–617.

Wright, T. A., & Cropanzano, R. (2000). Psychological well-being and job satisfaction as predictors of job performance. *Journal of Occupational Health Psychology, 5*(1), 84–94.

Yeoman, I., Robertson, M., Ali-Knight, J., Drummond, S., & McMahon-Beattie, U. (Eds.). (2012). *Festival and Events Management* (1st ed.). London: Routledge.

Zehrer, A., Muskat, B., & Muskat, M. (2014). Services research in tourism: Advocating the integration of the supplier side. *Journal of Vacation Marketing, 20*(4), 353–363.

4 User-generated events in tourism

A new must-not-miss phenomenon

Estela Marine-Roig, Eva Martin-Fuentes and Natalia Daries-Ramón

Introduction

A decade ago, something unprecedented and unexpected occurred when an Australian teenager's invitation to a party via the social network MySpace went viral and more than 500 people turned up, resulting in damage to the neighbourhood and serious public disorder with police intervening; this event later inspired the film *Project X* (Van Dijck & Poell, 2013). Other similar spontaneous events generated by users through social media followed and became increasingly massive, exemplifying the convening power of social media and its potential to go viral (Van Dijck & Poell, 2013; Zanger, 2014).

Since then, events completely generated by users and promoted through social media have not only been growing, but have also become more organised, formal and specialised in certain topics (Lee & Tyrrell, 2012; Shirky, 2011) such as tourism. These tourist events have a great capacity to attract participants, generate income for tourist firms and destinations and promote tourist brands. In this respect, social media have not only revolutionised the way we communicate and publicise in travel and tourism (Hays, Page, & Buhalis, 2013) but also changed the role of tourists in the travel process, from relatively passive agents to empowered agents actively generating content and engaging with one another (Sigala, Christou, & Gretzel, 2012). In this respect, special attention is applied to user-generated content (UGC) online. This is due to the influence it has on destination and business image formation, tourist behaviour, and tourist satisfaction (Kim & Stepchenkova, 2015; Marine-Roig & Anton Clavé, 2016; Martin-Fuentes, 2016; Martin-Fuentes, Fernandez, Mateu, & Marine-Roig, 2018; Sparks, Perkins, & Buckley, 2013; Xiang & Gretzel, 2010), as well as its importance for tourism research and hospitality applications (Lu & Stepchenkova, 2015). However, the great capacity of social media to organise and publicise events has been acknowledged (Getz, 1997; Hudson, Roth, Madden, & Hudson, 2015; Montanari, Scapolan, & Codeluppi, 2013), as has its use by tourism agents, especially destination management organisations (DMOs) to publicise events (Hays et al., 2013; Sigala et al., 2012). Little attention has been given to user-generated events (UGEs) in tourism, which are events with travel and tourism purposes created, organised, held, and promoted entirely by users (rather than tourism organisations or businesses) through social media, closely bound to UGC and online communities.

Hence, the aim of this study is to define the concept of UGEs through social media in the field of event tourism by identifying their specificities around the new empowered role of users, and to assess their promotional capacity in relation to social media. Baring this in mind, this chapter aims to provide a quantitative framework approach to storing, analysing, and comparing events in relation to social media, which will be complemented with qualitative, observant participation at different events to deepen the understanding of user profiles, behaviour, status, motivations, and faithfulness. This mixed-methodological framework underpins the study of different Instagram Meetups at Pyrenean ski resorts.

Theoretical background

The socio-economic and socio-political turbulence, as well as the devastating changes in the natural environment, has led to the expansion of critical studies in the area of event management (Robertson, Ong, Lockstone-Binney, & Ali-Knight, 2018). Events cannot be explored or understood without understanding their social, cultural, and political context (Spracklen & Lamond, 2016). Thus, recently, critical event studies (CES) have proliferated and have been the core of publications such as books (Finkel, Sharp, & Sweeney, 2019; Lamond & Platt, 2016; Spracklen & Lamond, 2016), special issues of journals (Robertson et al., 2018), and scientific articles (Lamond & Agar, 2019; Lamond & Reid, 2017).

In line with matters developed by the previous CES, the critical study of an event needs to first raise two issues that do not have a unique answer in the literature on event management: what constitutes the event and what is the context in which it takes place? These questions are not separate issues: the second depends on how the first is understood (Spracklen & Lamond, 2016). In the following sections, we will try to answer these questions by defining the tourism events (meetups) convened and managed by the participants through social media. The social and cultural contexts are determined by the virtual community (Instagram), participants are members and by their individual involvement or engagement part of the online community.

Community-based events

In Arnstein's ladder of citizen participation (Arnstein, 1969), the highest degree of power is *citizen control*, which guarantees that residents or participants can govern a program, be fully in charge of the aspects of policy and performance management, and be able to negotiate the terms with third parties. Their level of engagement is very high because the community initiates, manages, and governs the project, and therefore the facilitators do not lead. In the context of the events, participants can choose which event they want and how they want to achieve it (Bostock, Cooper, & Roberts, 2016).

Today, the role of users at events has shifted from a passive one to an active one in which they co-produce and co-create the event; they provide ideas, meanings, and become brand ambassadors empowered by social media (Hartmann, 2012; Zanger, 2014). In this context of user empowerment, events completely generated by users have emerged and have not only been growing in number, but are becoming more formal, organised, and specialised in certain subjects (Lee & Tyrrell, 2012; Shirky, 2011).

Considering that a meeting is a generic term applicable to a gathering of people for any purpose (Getz, 1997), a meetup can be defined as an offline meeting of a physically dispersed virtual community (Sessions, 2010). That is, a meetup is an event set up by a virtual community and dedicated to the community itself. Meetup attendees strengthen ties to other attendees, improve the engagement to the online community as a whole, and contribute to the creation of bonding social capital (Sessions, 2010). There are even proliferating social networks such as "Meetup" to organise events by users, with 315,000 monthly meetups (Roy & Zeng, 2015), many of them eminently touristic.

Event tourism

The study of tourism events has increased dramatically since 2000, and subject areas have diversified (Kim, Boo, & Kim, 2013). Events generate tourism demand and several studies emphasise their importance for the development of a destination. The perceived image of a destination and the quality of an event are strongly related to visitor satisfaction and that both the image of the destination and of the event combined influence the decision to travel to that destination (Kim, Kang, & Kim, 2014; Lai & Li, 2014). Tourism-related events have been studied in relation to a wide range of topics (Getz & Page, 2016). Some of the key research aims in event tourism are to study how the level of engagement affects the tourism event experience and how the community spirit is shaped by events (Getz, 2008). Few studies have analysed events from the point of view of tourist promotion or marketing. Further, many events are not promoted from a tourist point of view at all (Hudson et al., 2015). Hence, social media are powerful tools to promote, commercialise, and shape the experience of an event (Bolan, 2014; Zanger, 2014) and they can help to attract tourists to the destination of the event (Getz, 1997) through electronic word-of-mouth (eWOM). This especially is interesting for destinations and companies, as Gretzel and Yoo (2008) showed that consumers' eWOM through social media generates more trust than communications from the company itself. In social networks, contacts (followers and followed) are members of the consumers' existing networks and may be perceived as more trustworthy and credible than unknown strangers (Chu & Kim, 2011).

Instagram as a promotion tool

In recent years, events promoted through social media based on photographs or images have grown with the proliferation of the use of smartphones and other mobile devices with built-in cameras. These new types of social media based on images have become increasingly important due to their usefulness and the ease with which users can share content, attracting a younger audience with a high return rate. In these media, each image has attached identification metadata through which users can search, navigate, and order according to their interests and priorities (Lister, 2011), and it can be accessed by millions of users worldwide. Instagram is the image-based social network that has become quite popular in recent years (Paül i Agustí, 2018). There were more than 400 million active monthly users, 30 billion stored pictures (the biggest *phototheque* on Earth), and 70 million photographs shared daily by January 2016 (Systrom, 2016), and surpassing one billion active monthly users and 500 million daily active stories by January 2019.

Instagram has become a reference medium to share creative experiences of people's lives and goes far deeper into emotions than other social media. Instagram can be effectively integrated in the publicity of brands or products thanks to its everyday character which humanises brands and allows users to participate in their daily scenes and processes, in a less institutional way, generating affect and engagement (Carah & Shaul, 2016). As these authors explain, being an Instagram user is about posting and circulating everyday experiences, "instantly", while they happen and where they happen (through geo-tagging), and also about engaging with live and ephemeral flows of images by scrolling, liking, and commenting: images receive most attention within the first several hours of being posted, and then mostly disappear from view. The credibility of Instagram's pictures may be questioned due to their hyperreal idealised nature (Borges-Rey, 2015) and the extended use of appealing filters (Erkan, 2015); it has been found that they are highly influential and trusted as they change users' perceptions and behavioural intentions at a pre-trip stage (Shuqair & Cragg, 2017), when filters are used to improve or add a personal touch to the picture, but not to alter reality (Thelander & Cassinger, 2017).

Analysing Instagram helps gain insights into social, cultural, and environmental issues about people's activities (through their camera lenses) (Hu, Manikonda, & Kambhampati, 2014; Paül i Agustí, 2018). Moreover, tourists who post pictures on social media differ in their purchasing behaviours from those who do not, as they are more likely to purchase souvenirs (gifts and products) as evidence of travel, and are more interested in purchasing regionally specific arts and crafts and special local products, such as food, than non-picture posters (Boley, Magnini, & Tuten, 2013). This especially is interesting for destinations and tourist companies as they potentially may generate larger economic impacts and boost local economies.

Especially remarkable are Instagram communities concerning certain topics (Ferrara, Interdonato, & Tagarelli, 2014), many of which revolve around tourist destinations by displaying pictures of them using specific hashtags to tag images of attractions, moments or places at the destination (such as #igers_place or #ig_place). In this respect, through Instagram, several social phenomena generate: photography competitions, courses, insta-talks, insta-walks, prize draws, and Instagram Meetups or events, many entirely generated and run by users. Besides, events and promotional campaigns generated through Instagram lead to a greater involvement of followers than other types of social media such as Facebook. Moreover, the post type events promote linear impact on likes and comments on Instagram (Coelho, Oliveira, & Almeida, 2016). However, UGEs in tourism are under studied as a distinct social phenomenon, especially linked to online communities and social media, despite their enormous promotional and attraction capacities and important implications for tourism companies and destinations.

Methodology

Possible research methods in event tourism are: hermeneutics, phenomenology, direct and participant observation, and experiential sampling (Getz, 2008). The combination of qualitative and quantitative techniques is used to achieve a wider understanding of the issue under study (Walle, 1997). In this case, to achieve a more holistic understanding of the phenomenon of UGEs, this combination of methods specifically suits our study. This study adopts a mixed-methodology to analyse and characterise the phenomenon of tourism UGEs. It includes a proposal of a quantitative analysis framework to store, analyse, and compare an event in relation to social media and its promotional effect. In addition, people actively participate as social media users, along with a qualitative approach to complement the quantitative analysis based on observant participation in the events in order to gain deeper understandings about social media users' opinions, behaviours, and experiences. This method frames the case study of three Instagram Meetups at ski resorts.

Analysis framework (quantitative approach)

Previous studies have determined the capacity to disseminate tourist image or brand and level of user engagement with UGC posts through the analysis of interactivity, reactions (likes, comments, and shares), and searches generated in social media (Huertas & Marine-Roig, 2015, 2016, 2018). Specifically, engagement with Instagram posts has been analysed through followers, comments, and likes (Abbott, Donaghey, Hare, & Hopkins, 2013; Carah & Shaul, 2016). Moreover, the "buzz" or dissemination of events through social media has been analysed through uploads of posts, such as photographs and videos, and the number of views they have had; the role of "influencers" is especially emphasised (Hartmann, 2012). For example, there are instagrammers like Selena Gomez (@selenagomez) and Ariana Grande (@arianagrande) who have about 150 million followers at present.

From a quantitative perspective, to store, analyse, and compare UGEs through social media a model which can be used and adapted to any type of event and/or social media platform is proposed. It is based on identification and analysis of the UGC posts produced in relation to the event, in order to obtain information about the participants, promotion, and the event itself.

Figure 4.1 represents a simplified class diagram to store and analyse an event. Boxes have three sections: Class name, Attributes and Methods. The Event class is associated to the User class and with the Firm/DMO class with a cardinality 1..*, that is to say, an event is associated to an indefinite number of users and firms and, therefore, we can know in every moment how many users and companies participate. User class is associated with an indefinite number of photographs (posts). The diagram shows the basic attributes gathered per class: time, likes, and comments per post; profile (gender, for individuals, and type of organisation for other users) and followers per user; event name, hashtag, location, date, and influx; and name, branch, and role per firm/DMO. Moreover, an event has five basic methods that go through the list of users and through the photograph list of each user:

- Photo(): The total quantity of pictures submitted with the corresponding hashtags.
- Dissemination(): The quantity of potential Instagrammers who have received the pictures in first instance, by adding up the followers of users.

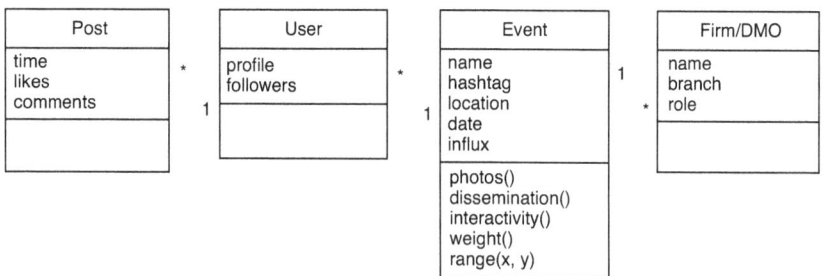

Figure 4.1 Unified modelling language (UML) diagram to store and process an event.
Note. cardinality 1: one entity; cardinality *: zero or more entities.
Source: Marine-Roig, Martin-Fuentes, and Daries-Ramon (2017).

- Interactivity(): The quantity of reactions (likes and comments) that users' posts had.

To be able to compare different events, the Event class counts on auxiliary methods:

- Weight(): The growth in percentage that the number of user participants represents over the average visitors (influx) when there is no special event. This enables the comparison of event impact in different areas and/or dates.
- Range(x, y): Different types of ranges such as posts per user (to detect the most active users), followers per user (to detect their role as "influencers"), likes per post, and comments per post (to detect the most engaging posts). Ranges are calculated in percentage.

The Firm/DMO class and the association with the Event class enable knowing the quantity, name, branch, and the role they play in relation to the event (support, sponsoring...). With this model, the different attributes and methods can be aggregated and compared within an event or across events. The number of repeat or new participants at different similar events should be identified to detect faithfulness.

Participant observation

Participant observation is a recognised method in events and tourism research, with special utility in areas such as tourist satisfaction/dissatisfaction, behaviour and management questions (Bowen, 2002), as well as in the observation of tourist groups or events (Markwell, 1997). Specifically, it has been used for the analysis of tourist behaviour in relation to photography (Gillet, Schmitz, & Mitas, 2016; Markwell, 1997) and online photo-sharing (Konijn, Slumer, & Mitas, 2016). This technique is considered as being microethnographic (Konijn et al., 2016), which analyses social and cultural interactions in an organised context or particular situation.

Participant observation is especially appropriate for exploratory and descriptive studies (Jorgensen, 1989), providing observations to help the researcher better understand the context and the phenomenon of study (DeWalt & DeWalt, 2010). To gain a better understanding of UGEs (Instagram Meetups), participant observation aims at observing the tourists' behaviours and interactions. The specific application of participant observation is explained in the case study section.

Case study: meetups of Instagrammers in the Pyrenees

The Instagram online community DescobreixCatalunya (DiscoverCatalonia) organised three different "Instagram Meetups" at three ski resorts in the

Eastern Pyrenees: La Molina (LM), Font Romeu (FR), and Les Angles (LA) during the months of January and March 2017 and January 2018 (Table 4.1). The DescobreixCatalunya community, with the profile @DescobreixCatalunya, was created in October 2012 by a group of friends interested in the tourism and promotion of Catalonia as a tourist destination, especially for the local public.

This community had previously agreed with the ski resorts to give discounts to the participants in the event for ski passes and equipment rental. The Instagram Meetups were described as leisure and discovery events in relation to winter activities (all offered plenty of leisure activities in addition to skiing such as adventure sports, nature trips, or snowshoeing trips) and were directed at the public in general, especially families. The participants had to be Instagrammers and follow the ski resort's Instagram account to be able to get the discounts and participate in the photography competition. Registrations to the events were limited to members of the Instagram community, and one accompanying person per Instagrammer.

Participants, usually active Instagram users, were motivated to post pictures on Instagram with the corresponding hashtags (see Table 4.1) on the day of the event or the days immediately after because they would participate in a competition. To enter the contest, photographs were geotagged, taken on the day of the event, have the hashtags of the event, and be posted in a public profile. There were five prizes for the best photographs (e.g. an adventure pack for two or a spa pack).

Hence, for *quantitative analysis*, all photographs with the hashtags proposed by the community @DescobreixCatalunya used to label the pictures

Table 4.1 Meetups in the Eastern Pyrenees

	La Molina	*Font Romeu*	*Les Angles*
Ski area/ slopes	61km/53	43km/41	55km/45
Winter season visitors	330,491	455,000	380,000
Approx. daily visitors	2562	3760	2815
Day of the event	29/01/2017	05/03/2017	14/01/2018
Participant places	200 places	250 places	200 places
Specific event hashtags	#descobreixlamolina	#descobreixfontromeu	#descobreixlesangles
Instagram account	@lamolinaski	@fontromeu.p2000. altiservice	@lesangles_ levillagestation

taken during the events, posted on the day of the event, and six following days were gathered and analysed. This period of photo analysis is consistent with Instagram usage patterns. Pictures are posted on the web shortly after being taken. At times old pictures are used, but not usually or socially accepted within this network, as there are even some specific hashtags asking for "permission" to post old pictures such as #TBT or #ThrowbackThursday. The proposed quantitative model underpinned three events of the case study, and different attributes for Posts, Users, the Event, and the related Firm/DMO were gathered and the methods applied.

For the *qualitative analysis* of UGEs through participant observation, guiding principles as set out by Bowen (2002) were applied: the research problem was approached from the perspective of the tourist (insider), the events were not created or manipulated by the researcher, and the setting was natural. The researcher was constantly aware of the need to forge a balance between a passive and a proactive presence. During the observant participation, DeWalt and DeWalt's (2010) suggestions were followed: actively observe, attending to details; look at the interactions, counting persons, or incidents; listen carefully to conversations and nonverbal expressions to seek out new insights; and keep a running observation record.

In this case, observation was performed by a team of two researchers in each of the three settings (LM, FR, and LA), between 7am and 4:30pm. The events had a predetermined size limit with regard to user numbers (see Table 4.1). According to Bowen's (2002) basic typology of field roles, the researchers were full participants, registered for the events. The observed group consisted of participants at the event performing different activities and researchers joined groups of Instagrammers for meals or breaks.

Results

This section first presents the quantitative results obtained through the application of the proposed analysis model, and then provides the qualitative results of participant observation.

Quantitative analysis

Participants posting pictures on Instagram and dissemination

The first aspect we can observe is that on the day of and six days after the events 50–90 different users posted photographs in each event. These users had an average of around 8,000 followers and posted about 2.5 photographs each. Remarkably, the numbers of users, followers, and average posts were extremely similar and proportional across the three events (Table 4.2), which may give an indication of the impact of similar future events. Using the Dissemination() calculation, we can see that these users potentially ranged in total from 373,787 followers, in the case of LA, to 738,800, in the case

Table 4.2 Information about Instagram users who posted photographs

	Different users posting photos	Total followers of users	Average followers per user	Max. followers	Average posts per user	Max. no. of posts
LM	88	738,800	8,397.89	126,000	2.71	13
FR	52	465,894	8,959.50	126,000	2.50	7
LA	50	373,787	7,475.74	126,000	2.40	11

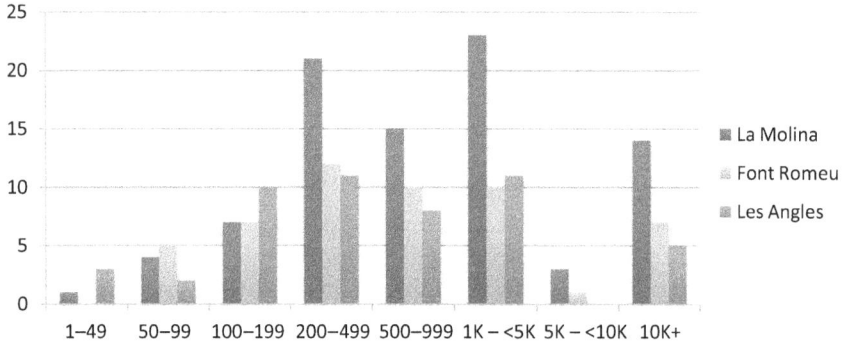

Figure 4.2 Followers per user.

of LM. This is the maximum threshold of dissemination of about half a million followers at each event.

Using the range(x, y) method, with regard to followers per user (Figure 4.2), most frequently, users had between 200 and 499 followers or between 1,000 and 5,000 followers, which is a significant number of potential people who may see each of the posts. The next largest cluster for LM and FR was the one with 500–999 followers, which can generate high image dissemination and interaction. The most influential users had more than 10,000 followers each and participated in multiple meetups. This target segment is especially interesting for marketing purposes for tourist destinations and businesses. The community @DescobreixCatalunya also posted photographs and had more than 126,000 followers.

Using the range(x, y) method, posts per user were identified (Figure 4.3): in general we can see that most users posted only one photograph, followed by users who posted three to four pictures or two pictures. Rarely do users (about 2%) post more than ten pictures. This can be explained by the subculture of the Instagram users who tend to post the "PicOfTheDay" and select only the very best pictures.

Instagram user participants who posted pictures related to the events were mostly individuals, more male than female (Figure 4.4). All three ski

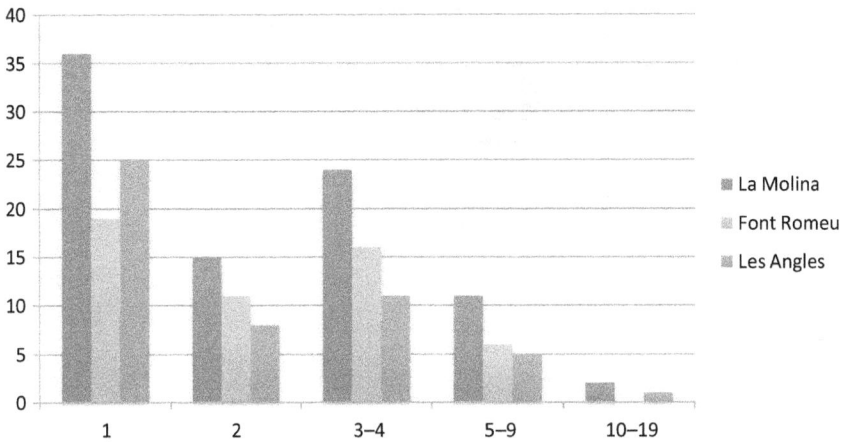

Figure 4.3 Posts per user.

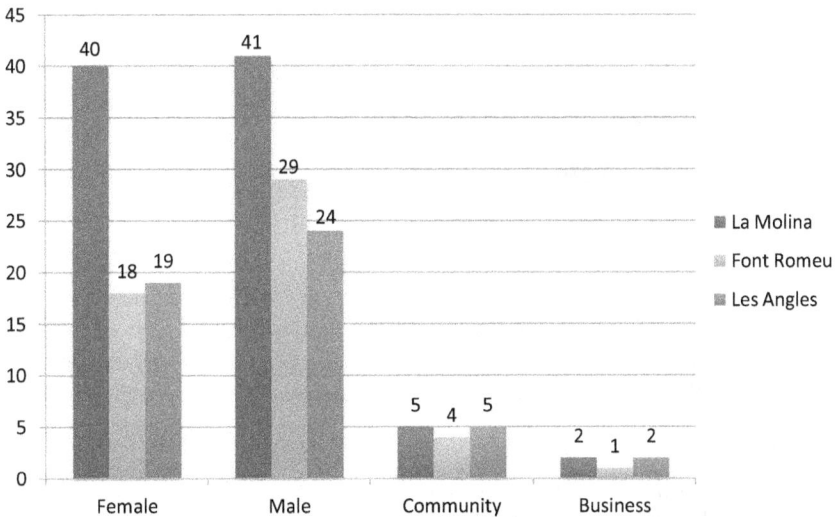

Figure 4.4 User profile.

resort profiles (@lamolinaski, @fontromeu.p2000.altiservice, @lesangles_levillagestation) actively participated in the event by posting pictures.

In terms of user faithfulness to these events (Table 4.3), in adding up the three consecutive events, 227 different Instagram users participated and posted pictures and of these, remarkably, 16.3% of users attended more than one event, and about 3% attended all three events. This indicates a moderate

Table 4.3 Repeat participants to the events

	LM	FR	LA	LM+FR	FR+LA	LM+LA	All 3	Sum
No. of users	88	52	50	17	5	8	7	227
Per cent	38.77	22.91	22.03	7.49	2.20	3.52	3.08	100

Table 4.4 Information by photographs posted related to the events

	Total photos	Total likes	Average likes per photo	Max. likes	Total comments	Average comments per photo	Max. comments
LM	238	33,983	142.79	10,700	1,388	5.83	118
FR	130	12,237	94.13	1,565	800	6.15	137
LA	120	9,640	80.33	1,339	435	3.63	75

degree of faithfulness to the Instagram community and to a certain type of events, while at the same time it shows the events have the capacity to attract new people. More than 15% of repeaters were very significant as the events were held on different periods and in relatively distant locations, indicating generally positive satisfaction with the experience.

Posts and interactivity through Instagram

With regard to photograph posts that had both hashtags related to each of the events (Photo() method), each event generated between 120 and 238 photographs (Table 4.4), with about 10,000–34,000 likes (Interactivity() method). Each event generated between 400 and more than 1,300 comments (Interactivity() method). This shows interactivity with posts, where multiple users, including the participants at the event, actively and purposefully liked or commented on the photographs. Moreover, each post had between 80 and 142 likes on average and between three and six comments, and the most popular post which belongs to the community @DescobreixCatalunya had more than 10,700 likes and 137 comments alone.

Using the range(x, y) method to identify likes and comments per post, we see that for two of the events (FR and LA) about 30% of posts had less than 50 likes (Figure 4.5). In the case of LM, however, this figure is only 20%. Usually there are fewer and fewer posts as likes increase. However, it is not negligible that about 40–50% of posts had between 50 and 199 likes, and that 21% of posts, in the case of LM, had between 200 and 499 likes. Remarkably, more than 10% of posts in LM had more than 1,000 likes, which entails very high interactivity. The two latest events (FR and LA) seem to behave much more similarly than the first (LM) in terms of likes.

Comments (Figure 4.6) were less frequent at the last event (LA). At the three events, the most frequent case is that pictures had one or no

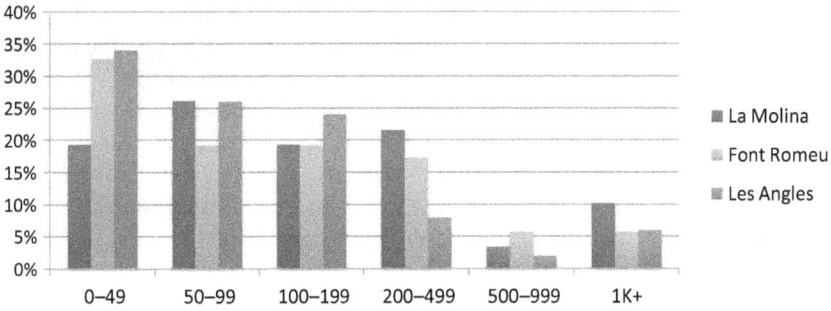

Figure 4.5 Range of likes per post (%).

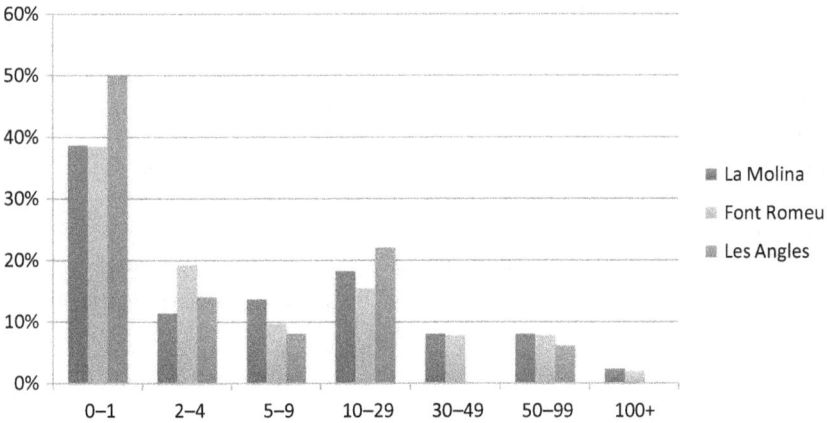

Figure 4.6 Range of comments per post (%).

comments, and then the pictures had between 10 and 19 comments, indicating higher interactivity. Again, some specific pictures ranked highest (surpassing 100 comments), usually corresponding to the users with the greatest number of followers and generating true conversations around the event or picture.

Along *overall lines*, we can assert that the three events behaved quite similarly although the event of LM received more visitors with more followers and posted more pictures in general and on average, and registered a considerably higher number of likes and comments in absolute numbers. Moreover, the top segments of the graphs (most influential) were higher in the case of LM. Arguably, the first event enjoyed the novelty factor to attract more interest and interactions. Probably, at first only the most engaged social media users with the Instagram community and that have high numbers of followers knew about and participated in it, whilst after the success and dissemination of the first event these events reached a broader public that may not be so close to the online community. In relative terms, the impact of participant numbers for the ski stations, applying the Weight() method, was notable even if the number of participants at the events was strictly limited: for LM, the event participants represent an increment of 7.8% on their average daily visitors. For FR, this represents an increase of 6.7% in daily visitors and of 7.1% for LA. Hence, we see the special interest UGEs may have for small ski resorts (destinations or companies).

Qualitative analysis

Participant observation

Participant profile: Most participants were 30–45 years old, although some younger in their twenties, and in general had a medium-high education level and medium purchasing power. Some participants, about one in four, came with their children. Most did not work in jobs related to photography; some mentioned being teachers, secretaries, engineers, business men/women. From what we could see and hear, because of the language spoken, the large majority of participants were Catalan, with large numbers from Barcelona.

Faithfulness and sense of belonging to the online community: all participants the researcher talked to or heard in conversation expressed their will to attend other similar meetup events. Some people argued that participating in the event was a cheaper way to visit places because of the discounts. In conversation, many participants expressed that the event was fun, that they were happy to have come, and that they were enjoying the activities and the company of other users.

Basic travel motivators: Physical motivation in participants occurred in both sport/exercise and relaxing entertainment. Although most people participated in skiing and other sports activities, some remained in the cafeterias or in leisure areas and many stopped regularly for a drink, or even to play cards with friends or family. From what we could see, most participants had already skied before or were regular skiers, demonstrating how this type of event attracts a specific segment of the community interested in this kind of sport. With regard to prestige/status motivations, most users expressed that they had used Instagram for some time and that they were active users. This shows how this social media can play an important part in their social status. In terms of leaders and hierarchy, usually most of the high "influencer" users with thousands of followers were part of the organisation and wore an identification badge. Therefore, people went to them for information and, hence, gained a higher status because they proved themselves most knowledgeable and involved in the events.

Behaviour and relationship to photography and social media: During the events, although some participants used cameras, most of them used smartphones to take multiple pictures and upload them to social media, and to participate in the event competition. Interestingly, because the Instagram photography competition implied originality, the staging of the pictures was highly accentuated, to set the frame before taking a "perfect" picture. Tourist photography is seen as an acquisitive or even predatory act (Markwell, 1997). In this respect, early in the morning of the events, many participants were taking pictures of the best views of the sight, trying to be the first ones to post images of these places on Instagram with the event hashtags. Many users acted somewhat surreptitiously, trying to find a unique view or spot to take pictures, but some, especially in group activities, adopted an imitative attitude and copied others' ideas for pictures by taking similar shots of the same things. However, in spite of the photography competition, many participants said that taking and posting pictures on Instagram was a daily activity they did to share experiences with certain online communities, friends and family, and for fun.

Discussion

The results showed that tourist UGEs are intrinsically related to the production of online UGC through social media and generate user engagement and interactivity through likes and comments. Social media are integrated within UGEs, to the extent that the event would not be possible, would lose all its significance or would not have happened without the specific social media, and would not have any significance either without people travelling physically to the site.

In addition, the promotional capacity of UGEs through online image/brand dissemination is extremely important. About 200 Instagram users posted 500 pictures, creating high engagement and online "buzz", receiving more than 55,000 likes and more than 2,600 comments in total, and potentially reaching more than one and half million people (followers). This daily activity happens with little or no cost to destinations or tourism businesses – both in terms of time and money. This especially is relevant as social media users who post pictures online potentially spend more money and consume more at the destination than those who do not (Boley et al., 2013).

Remarkably, along the lines of Hartmann (2012), a few highly influential users were detected, who accounted for the majority of online dissemination. Because of the fact that they originated from an online community, tourism UGEs may entail greater dissemination than regular events that use social media as part of their promotion. At UGEs, and especially Instagram Meetups, UGC production is part of what people "have" to and "want" to do to participate in the event. Participants partly engaged in the photography competition to win a prize. Nevertheless, if the prize had not existed, they probably would have posted pictures of the event anyway, for community recognition to gain a reputation or prestige as an Instagrammer, and for the hedonism of posting experiential pictures, as most participants were active users, many posting daily pictures of their experiences.

In UGEs, the reproduction of tourist images (Markwell, 1997) and the closing of the hermeneutic circle (Caton & Almeida Santos, 2008; Marine-Roig, 2019) through social media may be more intense, conscious, and purposeful, and contribute highly to promoting the destination or organisation. Analysis shows that only a few selected photographs were posted on Instagram, probably the pictures users consider the "best" or most "original", offering an ideal picture of the place or the event. UGEs can also help to reinforce destination brand identity and image, as well as create and reinforce territorial links, related to tourism assets and offers.

Furthermore, as shown in the quantitative results, UGEs have a great capacity to attract people, and specifically Instagram UGEs can have a great deal of influence on and attraction for destinations, reinforcing the idea of Baksi (2016) of their emotional bonding capacity, their ability to generate favourable images of a destination, and loyalty. Notably, manifest content of online photographs has been found to influence the attitudes towards a destination and affect destination image formation (Kim & Stepchenkova, 2015). If UGEs are successful and participants enjoy them and have fun, as the observant participation indicates, they will attach and transmit these sensations and feelings in relation to the destination, reinforced by followers' comments. The positive link between taking photographs and tourist satisfaction and happiness (Gillet et al., 2016) is thus supported by observant participation. These positive sensations can then be a source of inspiration for other travellers, increasing their desire to travel to the destination and participate in future events.

On top of that, these types of events organised by peers of the tourist and the information they transmit through eWOM to their online communities can have a greater impact and provide more effective and trustworthy marketing for the destination or organisation than official sources, as argued by several studies (Chu & Kim, 2011; Gretzel & Yoo, 2008). Official sources are seen as traditional forms of advertising (overt induced sources) with vested interests (Chu & Kim, 2011), while eWOM can be considered an organic source and the most trustworthy according to Gartner's (1993) classification of tourist information sources. This idea is supported by observant participation, which showed users' attachments and identifications with the community. Particularly, Instagram UGEs may be highly influential and capable of changing users' perceptions and behaviours in relation to a trip or destination (Shuqair & Cragg, 2017).

Tourism UGEs, especially in the case of Instagram Meetups, and unlike other spontaneous events created through social media, show the organisational capacity and empowerment of users as they are fully structured or semi-structured, organised and managed by users, and may entail formal registrations and membership. Our results show that ski resorts and business brands participated and mixed with participants as just another participant, informally and not as "organisers", thus becoming more "human" in the eyes of tourists. Their brand was clearly integrated in the event through the photography competition.

Previous studies (Hartmann, 2012; Zanger, 2014), acknowledging people's new empowered and central role in events as co-creators through UGC, always assumed that the company or the destination is a co-organiser. Further, they have an important degree of control over the event, or that social media integration to the event is at the initiative of the organisation and part of a marketing strategy. Nevertheless, the empowered role of users in events, enabled by social media, can be much deeper, to the point of a complete paradigm shift in which users are the main initiators, creators, organisers, and realisers of the event. As pointed out earlier, social media are useful tools not only to promote an event but also to organise collective action (Segerberg & Bennett, 2011). The shift is that the event is at the initiative of the people, not of the tourism company or tourist destination, and taken in a user milieu.

This reinforces the arguments which assess the empowerment of tourists and their increasing role as active agents (Hvass & Munar, 2012; Munar, 2011; Sigala et al., 2012) generating online content and events in a solely user milieu, for which tourism organisations have difficulty to access and have much less control. In this context, companies and destinations acquire a facilitating role, which may be more or less proactive and may entail promotion, sponsorships, but the initiative is not theirs. Therefore, the sentence "We need to learn to let go" (Lafley, as cited in Hartmann, 2012) may even acquire a deeper meaning or be used to go further, as the control of the event may be largely or even completely out of the hands of the destination or organisation. The concept could be transformed to: sometimes, tourism organisations need to let go and let customers take the reins!

Concluding remarks

Tourist UGEs can be defined as events completely initiated and organised by users through social media for tourism purposes; they have a certain structure and limits, are intrinsically bound to social media, and are the production of UGC and online communities. UGEs are currently a growing phenomenon and should be recognised and denominated as a differentiated type of event bound to social media, just as tourists' UGC are recognised as a differentiated type of content and research matter. UGEs are but an expression of the empowerment tourists have gained with the emergence of social media.

This study's approach to the phenomenon of UGEs has demonstrated their great capacity for attraction and promotion through the posting of related UGC, which is potentially much greater than for other types of events, with their potential to create user interactivity and the importance of "influencers" in the phenomenon. It has also shown their unique characteristics as a social phenomenon related to users' new empowered role as organisers and initiative takers, including high faithfulness and commitment of users, UGEs' ability to create emotional bonds with the brands, a strong sense of belonging to the community, and motivations related to social prestige and interpersonal relationships. Photography-based UGEs organised through social media, such as Instagram Meetups, are especially interesting for tourism destinations and businesses because of the intrinsic link between photography and tourism, and due to the projection of the "best" or most "original" pictures of a place with high doses of creativity, as well as for their specific capacity to transmit emotions and create engagement.

At a theoretical level, this study has contributed in defining and characterising UGEs in tourism. Although some previous event management and classification models (Zanger, 2014) have introduced the role of social media in them, future theoretical models should also incorporate the role of users in generating them: from total followers, to co-creators/organisers, up to full initiators/organisers.

At a methodological level, the proposed quantitative framework approach to store, analyse, and compare events based on Unified Modelling Language (UML) has enabled the processing of UGEs in relation to UGC and social media and can be used for the analysis of any type of event, both user and non-user-generated, and adapted to computerised analyses. Its combination with participant observation has been effective to gain a deeper understanding of UGEs and define their main characteristics, reinforcing the strength of mixed-methods to approach complex social phenomena.

Concerning managerial implications, tourist businesses and destination marketers should be vigilant and aware of the existence and importance of UGEs, of their potentialities for brand promotion and consolidation, and the creation of engagement and relationships with users, at very little or no cost to them. Although these events are fully generated by users, both DMOs and tourism destinations and organisations should try to play a role in them.

These types of events imply a paradigm shift in event management in that tourism organisations are no longer the main initiators and controllers of the event, adopting a secondary role of accompaniment and backing, but they need to take advantage of the fact that these tourist events usually need sponsorship, facilitation, or logistic and economic support. They should also take advantage of the capacity to humanise the brands of these events, and participate and mix with users. With UGEs, tourism destinations and companies need to learn to let go and let tourists take the reins, but they need to be on hand for support by providing attractive advantages for their development. If a destination or organisation is capable of attracting tourism UGEs, it may have found a holy grail, because people themselves will organise events for the organisation, at minimal cost and promote it massively through UGC and eWOM.

Concerning limitations and future research, this study analysed three meetups promoted by one specific online Instagram community in the same geographical area, so the results are only valid for this study. Replicating the research in different destinations with different online communities is a challenge for further research. The aim of this research is not analysing the economic impact of the events. Future studies should analyse other UGEs held through Instagram and other social media, and determine the different economic and promotional impacts and the return of the sponsors' investment. Moreover, it would be interesting to have a comparison of these events with other types of events organised directly by companies or institutions.

Acknowledgements

This work was supported by the Spanish Ministry of Economy, Industry and Competitiveness [Grant id.: TURCOLAB ECO2017-88984-R and TIN2015-71799-C2-2-P].

References

Abbott, W., Donaghey, J., Hare, J., & Hopkins, P. J. (2013). An Instagram is worth a thousand words: an industry panel and audience Q&A. *Library Hi Tech News, 30*(7), 1–6. doi: 10.1108/LHTN-08-2013-0047

Arnstein, S. R. (1969). A ladder of citizen participation. *Journal of the American Institute of Planners, 35*(4), 216–224. doi: 10.1080/01944366908977225

Baksi, A. K. (2016). Destination bonding: Hybrid cognition using Instagram. *Management Science Letters, 6*(1), 31–46. doi: 10.5267/j.msl.2015.12.001

Bolan, P. (2014). A perspective on the near future: Mobilizing events and social media. In I. Yeoman, R. Robertson, U. McMahon-Beattie, K. A. Smith, & E. Backer (Eds.), *The Future of Events and Festivals* (pp. 199–209). Oxford: Routledge.

Boley, B. B., Magnini, V. P., & Tuten, T. L. (2013). Social media picture posting and souvenir purchasing behavior: Some initial findings. *Tourism Management, 37,* 27–30. doi: 10.1016/j.tourman.2012.11.020

Borges-Rey, E. (2015). News images on Instagram: The paradox of authenticity in hyperreal photo reportage. *Digital Journalism, 3*(4), 571–593. doi: 10.1080/21670811.2015.1034526

Bostock, J., Cooper, R., & Roberts, G. (2016). Rising to the challenge of sustainability: Community events by the community, for the community. In A. Jepson & A. Clarke (Eds.), *Managing and Developing Communities, Festivals and Events* (pp. 16–33). New York, NY: Palgrave Macmillan.

Bowen, D. (2002). Research through participant observation in tourism: A creative solution to the measurement of consumer satisfaction/dissatisfaction (CS/D) among tourists. *Journal of Travel Research, 41,* 4–14. doi: 10.1177/0047287502041001002

Carah, N., & Shaul, M. (2016). Brands and Instagram: Point, tap, swipe, glance. *Mobile Media & Communication, 4*(1), 69–84. doi: 10.1177/2050157915598180

Caton, K., & Almeida Santos, C. (2008). Closing the hermeneutic circle? Photographic encounters with the other. *Annals of Tourism Research, 35*(1), 7–26. doi: 10.1016/j.annals.2007.03.014

Chu, S.-C., & Kim, Y. (2011). Determinants of consumer engagement in electronic word-of-mouth (eWOM) in social networking sites. *International Journal of Advertising, 30*(1), 47–75. doi: 10.2501/IJA-30-1-047-075

Coelho, R. L. F., Oliveira, D. S. de, & Almeida, M. I. S. de. (2016). Does social media matter for post typology? Impact of post content on Facebook and Instagram metrics. *Online Information Review, 40*(4), 458–471. doi: 10.1108/OIR-06-2015-0176

DeWalt, K. M., & DeWalt, B. R. (2010). *Participant Observation: A Guide for Fieldworkers* (2nd ed.). Plymouth: AltaMira Press.

Erkan, I. (2015). Electronic word of mouth on Instagram: Customers' engagements with brands in different sectors. *International Journal of Management, Accounting and Economics, 2*(12), 1435–1444. doi: 10.1177/2050157915598180

Ferrara, E., Interdonato, R., & Tagarelli, A. (2014). Online popularity and topical interests through the lens of Instagram. In *ACM Hypertext and Social Media* (pp. 24–34). New York: ACM Digital Library. doi: 10.1145/2631775.2631808

Finkel, R., Sharp, B., & Sweeney, M. (Eds.). (2019). *Accessibility, Inclusion, and Diversity in Critical Event Studies*. New York: Routledge.

Gartner, W. C. (1993). Image formation process. *Journal of Travel & Tourism Marketing, 2*(2–3), 191–215. doi: 10.1300/J073v02n02_12

Getz, D. (1997). *Event Management & Event Tourism*. New York: Cognizant Communication Corporation.

Getz, D. (2008). Event tourism: Definition, evolution, and research. *Tourism Management, 29*(3), 403–428. doi: 10.1016/j.tourman.2007.07.017

Getz, D., & Page, S. J. (2016). Progress and prospects for event tourism research. *Tourism Management, 52*, 593–631. doi: 10.1016/j.tourman.2015.03.007

Gillet, S., Schmitz, P., & Mitas, O. (2016). The snap-happy tourist: The effects of photographing behavior on tourists' happiness. *Journal of Hospitality & Tourism Research, 40*(1), 37–57. doi: 10.1177/1096348013491606

Gretzel, U., & Yoo, K. H. (2008). Use and impact of online travel reviews. In P. O'Connor, W. Höpken, & U. Gretzel (Eds.), *Information and Communication Technologies in Tourism 2008* (pp. 35–46). Vienna: Springer Vienna. doi: 10.1007/978-3-211-77280-5_4

Hartmann, D. (2012). User generated events. In C. Zanger (Ed.), *Erfolg mit nachhaltigen Eventkonzepten* (pp. 23–36). Heidelberg, Germany: Gabler Verlag. doi: 10.1007/978-3-8349-6885-2_3

Hays, S., Page, S. J., & Buhalis, D. (2013). Social media as a destination marketing tool: Its use by national tourism organisations. *Current Issues in Tourism, 16*(3), 211–239. doi: 10.1080/13683500.2012.662215

Hu, Y., Manikonda, L., & Kambhampati, S. (2014). What we Instagram : A first analysis of Instagram photo content and user types. In *Eighth International AAAI Conference on Weblogs and Social Media (ICWSM 2014)* (pp. 595–598).

Hudson, S., Roth, M. S., Madden, T. J., & Hudson, R. (2015). The effects of social media on emotions, brand relationship quality, and word of mouth: An empirical study of music festival attendees. *Tourism Management, 47*, 68–76. doi: 10.1016/j.tourman.2014.09.001

Huertas, A., & Marine-Roig, E. (2015). Destination brand communication through the social media: What contents trigger most reactions of users? In I. Tussyadiah & A. Inversini (Eds.), *Information and Communication Technologies in Tourism* (pp. 295–308). Cham, Switzerland: Springer. doi: 10.1007/978-3-319-14343-9_22

Huertas, A., & Marine-Roig, E. (2016). Differential destination content communication strategies through multiple social media. In A. Inversini & R. Schegg (Eds.), *Information and Communication Technologies in Tourism 2016* (pp. 239–252). Cham: Springer. doi: 10.1007/978-3-319-28231-2_18

Huertas, A., & Marine-Roig, E. (2018). Searching and sharing of information in social networks during the different stages of a trip. *Cuadernos de Turismo, 42,* 185–212. doi: 10.6018/turismo.42.08

Hvass, K. A., & Munar, A. M. (2012). The takeoff of social media in tourism. *Journal of Vacation Marketing, 18*(2), 93–103. doi: 10.1177/1356766711435978

Jorgensen, D. L. (1989). *Participant Observation: A Methodology for Human Studies.* London: Sage Publications.

Kim, H., & Stepchenkova, S. (2015). Effect of tourist photographs on attitudes towards destination: Manifest and latent content. *Tourism Management, 49,* 29–41. doi: 10.1016/j.tourman.2015.02.004

Kim, J., Boo, S., & Kim, Y. (2013). Patterns and trends in event tourism study topics over 30 years. *International Journal of Event and Festival Management, 4*(1), 66–83. doi: 10.1108/17582951311307520

Kim, J., Kang, J. H., & Kim, Y.-K. (2014). Impact of mega sport events on destination image and country image. *Sport Marketing Quarterly, 23*(3), 161–175.

Konijn, E., Slumer, N., & Mitas, O. (2016). Click to share: Patterns in tourist photography and sharing. *International Journal of Tourism Research, 18*(6), 525–535. doi: 10.1002/jtr.2069

Lai, K., & Li, Y. (2014). Image impacts of planned special events: Literature review and research agenda. *Event Management, 18*(2), 111–126. doi: 10.3727/152599514X 13947236947347

Lamond, I. R., & Agar, L. (2019). Beyond the frame: Use of augmented screenings as a visual methodology in critical event studies. *Event Management, 23*(2), 269–278. doi: 10.3727/152599518X15403853721213

Lamond, I. R., & Platt, L. (Eds.). (2016). *Critical Event Studies: Approaches to Research.* London: Palgrave Macmillan. doi: 10.1057/978-1-137-52386-0

Lamond, I. R., & Reid, C. (2017). Bringing together political communication and critical event studies. In *The 2015 UK General Election and the 2016 EU Referendum* (pp. 3–16). Cham, Switzerland: Springer International Publishing. doi: 10.1007/978-3-319-54780-0_1

Lee, W., & Tyrrell, T. J. (2012). Arizona meeting planners' use of social networking media. In M. Sigala, E. Christou, & U. Gretzel (Eds.), *Social Media in Travel, Tourism and Hospitality: Theory, Practice and Cases* (pp. 121–132). Surrey: Ashgate Publishing.

Lister, M. (2011). Too many photographs? Photography as user generated content. *AdComunica: Revista Científica de Estrategias, Tendencias e Innovación En Comunicación, 2,* 25–41. doi: 10.6035/2174-0992.2011.2.3

Lu, W., & Stepchenkova, S. (2015). User-generated content as a research mode in tourism and hospitality applications: Topics, methods, and software. *Journal of Hospitality Marketing & Management, 24*(2), 119–154. doi: 10.1080/19368623.2014.907758

Marine-Roig, E. (2019). Destination image analytics through traveller-generated content. *Sustainability, 11*(12), article 3392. doi: 10.3390/su11123392

Marine-Roig, E., & Anton Clavé, S. (2016). Affective component of the destination image: A computerised analysis. In M. Kozak & N. Kozak (Eds.), *Destination Marketing: An International Perspective* (pp. 49–58). New York: Routledge.

Markwell, K. W. (1997). Dimensions of photography in a nature-based tour. *Annals of Tourism Research*, *24*(1), 131–155. doi: 10.1016/S0160-7383(96)00053-9

Martin-Fuentes, E. (2016). Are guests of the same opinion as the hotel star-rate classification system? *Journal of Hospitality and Tourism Management*, *29*, 126–134. doi: 10.1016/j.jhtm.2016.06.006

Martin-Fuentes, E., Fernandez, C., Mateu, C., & Marine-Roig, E. (2018). Modelling a grading scheme for peer-to-peer accommodation: Stars for Airbnb. *International Journal of Hospitality Management*, *69*, 75–83. doi: 10.1016/j.ijhm.2017.10.016

Montanari, F., Scapolan, A., & Codeluppi, E. (2013). Identity and social media in an art festival. In A. M. Munar, S. Gyimóthy, & L. Cai (Eds.), *Tourism Social Media: Transformations in Identity, Community and Culture* (pp. 207–225). Bingley: Emerald Group Publishing Limited.

Munar, A. M. (2011). Tourist-created content: Rethinking destination branding. *International Journal of Culture, Tourism and Hospitality Research*, *5*(3), 291–305. doi: 10.1108/17506181111156989

Paül i Agustí, D. (2018). Characterizing the location of tourist images in cities. Differences in user-generated images (Instagram), official tourist brochures and travel guides. *Annals of Tourism Research*, *73*, 103–115. doi: 10.1016/j.annals.2018.09.001

Robertson, M., Ong, F., Lockstone-Binney, L., & Ali-Knight, J. (2018). Critical event studies: Issues and perspectives. *Event Management*, *22*(6-Special issue), 865–1108. doi: 10.3727/152599518X15346132863193

Roy, S. D., & Zeng, W. (2015). Media on the Web. In *Social Multimedia Signals* (pp. 9–18). Cham: Springer International Publishing. doi: 10.1007/978-3-319-09117-4_2

Segerberg, A., & Bennett, W. L. (2011). Social Media and the organization of collective action: Using Twitter to explore the ecologies of two climate change protests. *The Communication Review*, *14*(3), 197–215. doi: 10.1080/10714421.2011.597250

Sessions, L. F. (2010). How offline gatherings affect online communities. *Information, Communication & Society*, *13*(3), 375–395. doi: 10.1080/13691180903468954

Shirky, C. (2011). The political power of social media. *Foreign Affairs*, *90*(1), 28–41.

Shuqair, S., & Cragg, P. (2017). The immediate impact of Instagram posts on changing the viewers' perceptions towards travel destinations. *Asia Pacific Journal of Advanced Business and Social Studies*, *3*(2), 1–12. doi: 10.25275/apjabssv3i2bus1

Sigala, M., Christou, E., & Gretzel, U. (Eds.). (2012). *Social Media in Travel, Tourism and Hospitality: Theory, Practice and Cases*. Surrey: Ashgate Publishing.

Sparks, B. A., Perkins, H. E., & Buckley, R. (2013). Online travel reviews as persuasive communication: The effects of content type, source, and certification logos on consumer behavior. *Tourism Management*, *39*, 1–9. doi: 10.1016/j.tourman.2013.03.007

Spracklen, K., & Lamond, I. R. (Eds.). (2016). *Critical Event Studies*. New York: Routledge.

Systrom, K. (2016). Boomerang Instagram. Retrieved January 31, 2016, from https://www.instagram.com/press/

Thelander, A., & Cassinger, C. (2017). Brand new images? Implications of Instagram photography for place branding. *Media and Communication*, *5*(4), 6–14. doi: 10.17645/mac.v5i4.1053

Van Dijck, J., & Poell, T. (2013). Understanding social media logic. *Media and Communication*, *1*(1), 2–14. doi: 10.17645/mac.v1i1.70

Walle, A. H. (1997). Quantitative versus qualitative tourism research. *Annals of Tourism Research*, *24*(3), 524–536. doi: 10.1016/S0160-7383(96)00055-2

Xiang, Z., & Gretzel, U. (2010). Role of social media in online travel information search. *Tourism Management, 31*(2), 179–188. doi: 10.1016/j.tourman.2009.02.016

Zanger, C. (2014). *Ein überblick zu events im zeitalter von Social Media [An Overview of Events in the Age of Social Media].* Berlin, Germany: Springer. doi: 10.1007/978-3-658-05771-8

5 Undocumented! Small events of rural hinterlands, South Africa

Unathi Sonwabile Henama and
Petrus Mfanampela Maphanga

Introduction

Today, tourism is one of the world's largest industries and a big business due to its sustained growth.

> Tourism is one of the leading and top growing sectors of the global economy. Furthermore, it is described as one of the world's leading and most vibrant economic activities and has become an important economic sector in many parts of the world.
>
> (Labuschangne and Saayman 2014: 2)

The sustained growth rate of the tourism industry is virtually impossible. "Growth in the sector has made tourism a major generator of employment" United Nations Conference on Trade and Development (2010: 6). According to Botha (2012), many countries consider tourism as a means to increase income, generate foreign currency, create employment and increase revenues from taxes. It is for this reason that almost all countries have jumped on the tourism bandwagon, increasing the sheer number of tourism destinations. Steyn and Spencer (2011) noted that the change in government in 1994 opened up the country's vast tourism potential.

> The development of tourism in South Africa began its upward trajectory at the end of apartheid and the first democratic elections in South Africa. South Africa was once again open for business and international tourism grew to explore this new destination.
>
> (Henama 2017: 5)

South Africa's economic fortunes had been associated with extraction industries such as gold. However, in the recent past, the mining industry has been experiencing a downward spiral in employment and contribution to the Gross Domestic Product (GDP) and has been replaced by the tourism industry.

Henama (2014) noted the tourism industry as an engine for growth for the economy as traditional sectors such as mining and agriculture had witnessed a declining share in employment and contribution to GDP. Today, tourism is regarded as the "'new gold' attracts more foreign exchange than gold mining" (Henama 2016a: 2). "South African cities have sought to promote tourism as a driver for urban economic development through support for an array of different kinds of tourism" (Rogerson and Rogerson 2014: 94). South Africa has become the "new kid on the block" for tourism consumption, experiencing rapid growth in tourism consumption. Tourism has emerged as a reliable industry to achieve the objectives of the Reconstruction and Development Plan (RDP) and other successive economic development policies of the state. Besides South Africa's successful democratic transition, the country is still faced by structural challenges such as high unemployment rate, highest inequality and intergenerational poverty. The vast majority of South Africans are poor and unemployment is above 27%, whilst economic growth rates have slowed down to less than 1% in a country experiencing a democratic dividend. The ability of the tourism industry to create labour intensive jobs has immense potential to alleviate the challenges faced by South Africa. Booysens and Visser (2010) opined that tourism is increasingly regarded as a viable avenue for local economic development. "Tourism destinations attracts tourism because of the positive economic impacts such as labour intensive jobs, tourism acting as a catalyst for other industries, the attraction of foreign exchange and foreign direct investment that comes with tourism" (Henama 2016b: 1).

Tourism is important for the economies of developed and developing nations, regarded as a panacea for all economic ills. "Unfortunately, in some urban centres, unrealistic expectations abound regarding the role that tourism might fulfil as an economic growth mechanism for a town" (Ferreira 2007: 191). "The growth of tourism in South Africa during the past ten years resulted in the development of various types of tourism such as: eco-tourism, cultural tourism, adventure tourism, business tourism, sports tourism and event tourism" (Strydom et al. 2006: 87). Department of Sports and Recreation (2012) noted that South Africa adopted the National Sports Tourism Strategy in 2012. This occurred after the hosting of a plethora of international sporting events in South Africa with little benefit to the ordinary South Africans. "Tourism can play a significant role in overcoming the many socio-economic challenges South Africa faces" (Ferreira 1999: 313). "Tourism can provide one possible mechanism to re-distribute wealth from the rich to the poor" (Spencley and Meyer 2012: 299). "Tourism is too often regarded as a panacea-an economic, social and environmental 'cure-all'" (Chok et al. 2007: 146). Ferreira (2007) noted that unfortunately there are unrealistic expectations regarding the role of tourism as an economic growth mechanism. The fundamental lesson to learn from South Africa's tourism industry is that one off mega events are of little benefit for developing countries, and that small-scale events that use local resources are more sustainable.

Muslim religious festival: Ramadan in the Bo Kaap, Cape Town

The gang-infested areas around Cape Town will call a temporary truce every year during the month of Ramadan, where Muslims converge for prayer around the Bo Kaap, the centre of the South Africa' s Muslim industry.

> Every Ramadan a truce is called, and the gangs put their guns away because many of the leaders identify as Muslims. As Eid gets nearer, the community starts to feel fear seeping in again, because the end of the hold months marks the end of the peace.

Ramadan is a Muslim religious festival that is ripe for promoting Muslim tourism towards the Bo Kaap specifically and Cape Town generally. According to Dziewanski (2015: 2)

> religion is also one of few social resources available in a community where opportunities for empowerment and employment are scarce. In this context, Islam provides not only a belief system but also a community and identity outside of gang life...Ceasefire's statistics further suggest the potential of religion – and specifically Islam – to interrupt violence.

The Bo Kaap is also known as the Malay Quarter; according to the South African Historical Organisation (2019) the influx of Malays into the quarter began in the 1830s when slavery was abolished and Malays established their homes there as a Muslim neighbourhood. Onishi (2016) noted that the residents of the Bo Kaap are descendants of slaves brought from other African nations, India, Indonesia and Malaysia known over the centuries as Cape Malays or Cape Muslims. The unique architecture of the houses characterized by colourful single storey houses with flat roofs, cobblestone streets, and just a stone's throw away from the Central Business District (CBD) and having some of the oldest mosques in South Africa are what makes the Bo Kaap unique.

Apartheid legislation that prohibited the mixing of races resulted in the preservation of the Bo Kaap and its culture, which is associated with Islam. "A visit to Cape Town cannot exclude a visit to the Muslim 'Malay Quarters', an almost exclusive Muslim residential area, where visitors will be forgiven for thinking that Cape Town is a Muslim city" (David 2015: 256). Islam is a popular religious in the Western Cape and has emerged as a glue in the community. The number of crime and gangsterism ceases during Ramadan, as a plethora of Muslims head for the Bo Kaap during Ramadan. This period is characterized by reduced incidents of crime in the Cape Flats and other residential areas of the Western Cape. This therefore means the Bo Kaap is a successful crisis-resistant tourism destination frequented by domestic tourists to participate in Ramadan. According to Kamish (2019: 1) "Ramadaan, which starts the week of 5 May 2019, is a month of fasting from sunrise to sunset for Muslims around the world. What makes Ramadan in Cape Town so special is the infusion with the culture of the Cape Malay community, the descendants of East Asian exiles and slaves who arrived in 1600s". Dangor (1997) noted that Dutch colonialism was responsible for the introduction of Islam in Cape Town with Muslims brought to Cape Town from diverse regions of the world. According to Dziewanski (2015: 2)

> every year during Ramadaan, you will see throngs of Muslims gathered on the grass in Mouille Point for what looks like a family picnic at sunset. When you see that, it usually signals that Eid, the celebration at the end of the fast, is possibly the next day. It is a jolly affair of family and friends breaking their fast and praying together. On this day "maan kykers" (moon sighters) look out for the moon that confirms the end of Ramadaan. Next time you see this join in, non-Muslims are welcome for "boeka" (breaking of the fast). Ramadaan is about giving to the less fortunate, and so every Muslim must pay "fitrah" during the holy month, which means giving either R40 or 3kg of rice as a form of charity. Families will also put together food parcels of food basics to hand out to the poor or NGOs. On the last day of Ramadaan, organisations like Nakhlistan, a non-profit organisation involved mainly in feeding schemes, with the help of volunteers, cook over 100 pots of food, holding up to 130 litres of food to feed the poor during Eid.

The celebration of Ramadan in the Bo Kaap presents itself as an opportunity for tourism consumption. "Islamic festivals such as the two Eids, the Prophet Mohamed's birthday and sighting of crescent at the end of Ramadan, are events that transform patches of the city into hubs for celebration-an experience of a living culture" (David 2015: 262). "Religious motivated tourism is probably as old as religious itself and is consequently the oldest type of tourism" (Zamani-Farahani and Eid 2016: 3). Religious tourism is a segment of the tourism market in which tourists travel to visit religious centres and events in various countries to satisfy their beliefs. Muslim tourism has been growing rapidly and has emerged as a lucrative market segment for exploitation by destinations. There are an estimated 121 million Muslim international travellers, which is expected to grow to 156 million by 2020 according to Umar (2017). "In 2016, according to data of the Global Muslim Travel Index (GMTI) about $155 billion were spend by Muslim travelers. This number represents about 13% of the total global travel expenses" (Haupt 2017: 1). "Islamic tourism is a form of religious tourism because tourists visit religious rituals, conferences and ceremonies at national, regional and internationals levels and in religious venues" (Namin 2013: 1256). "Meccas is perhaps the best-known world pilgrimage destination, but there are many others, i.e. Saint Peter's Basilica in the Vatican City, Varanasi in India, Lumbini in Nepal or Jerusalem" (Fourie et al. 2014: 4). Hamza et al. (2012) noted that at Islamic tourism destinations, tourists can count on feeling secure that there are only family-centred activities and halal compliant food. In addition, there is no alcohol policy and segregated facilities so that women can swim in women-only pools. Prayer rooms at major airports and public and retail places improve the Muslim prayer experience, facilitating the five obligatory daily prayers, each performed within a specific window of time.

Muslim and non-Muslim countries equally attract Muslim international travelers with high disposable income and large travel groups. Consequently, being "Muslim friendly" is a prerequisite to increase the number of Muslim tourists at a destination. North West University (2017) noted that most Muslim tourists hail from Saudi Arabia, Iran, the United Arab Emirates, Indonesia, Kuwait, Turkey, Nigeria, Malaysia, Qatar and Egypt.

> South Africa in like fashion has positioned itself to be a halal tourism hub and attraction despite that about 2% of Muslims constituting its total population. Local platforms such as StayHalala as well as the ever thriving halal market and halal certified restaurants has made South Africa a major spot for Muslim travellers worldwide.
>
> (Haupt 2017: 2)

"Although South Africa is considered a non-Organisation of Islamic Cooperation (OIC) destination. Cape Town is one of Africa's leading Halaal tourism destination" (TNT News 2017: 2). Lubbe (1987) explained the Muslim kings incarcerated at Robben Island such as Tuan Said in 1744, Hadjie Matarim and Tuan Guru buried on Signal Fill in 1807. This means that Robben Island is a major Muslim heritage site. Davids (2015) highlighted the shrine of Sayed Abdurahman Motura and incarceration of a political prisoner buried on the island. Robben Island is popular for incarcerating anti-apartheid prisoners from South Africa such as Nelson Mandela, Govan Mbeki, Walter Sisulu and Robert Sobukwe. The struggle for independence against apartheid is more closely associated with Robben Island, indicating that the Muslim resistance associated with Robben Island remains unexploited – a dream deferred.

Inferences

Destinations that seek to attract Muslim tourist need to ensure that they improve their "Muslim friendliness". The election of Donald Trump as the American President is an opportunity for benefit from the Muslim un-friendliness of the United States and it is an opportunity for South Africa generally and Cape Town specifically to benefit from more Muslim tourism arrivals.

This requires bigger marketing outlays to be undertaken to present Cape Town as a Muslim friendly destination, considering the Muslim heritage at Robben Island and the Bo-Kaap. This includes the promotion of the Bo Kaap as a destination for Muslim tourists both during and outside of Ramadan. This requires caution considering the limited parking facility in the area to minimize the negative impacts of overtourism, which major tourism destinations are grappling with. Ramadan in the Bo Kaap is a tourism product offering to Muslim tourists to promote tourism. As noted by David (2015) Cape Town with its vibrant Muslim community and rich Islamic heritage is recognized as the model for 'halal tourism' in the country. North West University (2017) furthermore noted that KwaZulu-Natal Province (with Durban) and the Western Cape Province (with Cape Town) are ideal Muslim tourist destinations because of the high Muslim contingent. The theoretical framework indicates that activities that promote tourism can lead to a decline in crime, allowing domestic and international tourists and excursionists to consume Muslim heritage and religious festivals in the Bo Kaap, in the context of Ramadan. The legacy of hosting Ramadan in the Bo Kaap is that it becomes an advocacy event to declare it as a heritage protection site, as a means of preventing gentrification. Gentrification has become a major issue facing the Bo Kaap, and heritage protection lobbied for several years was finally achieved in the first half of 2019.

Theoretical framework: crisis-resistant tourists (Ramadan in the Bo Kaap, Cape Town)

"The tourism industry is particularly vulnerable to external shocks such as wars, diseases, extreme weather conditions, elections, adverse publicity, terrorism attacks, transport incidents, pollution, earthquakes, volcanic eruptions, political events, strikes, electricity shortages, recessions and fluctuations in economic conditions" (George 2014: 36). Osland et al. (2017) noted that terrorism, pandemic diseases and other threatening events have recently heightened the sense of personal risk for tourists. The tourism industry tends to be highly sensitive to negative environmental factors (Huang et al. 2008). The occurrence of a crisis and disasters at a tourist destination is naturally supposed to lead to a decline in demand for tourism in that region; this is particularly prone in a world where terrorism has become the new normal. Hajubaba et al. (2015) noted that there is a new target market that is inherently more resistant to crises and there is evidence that it reduces the crisis-vulnerability of tourism businesses and destinations. Hajubaba and Dolnicar (2017: 1) stated that "some segments of tourists are less likely to cancel their travel plans to a risky destination. Crisis-resistant tourists do not cancel their travel plans to a crisis-stricken destination and book despite crisis at the destination". The cancelation of travel plans to a crisis hit destination can have a detrimental impact on the economic fortunes of tourism destination, as tourists may consider their personal safety and security when making such decisions. "When a disaster happens, various aspects of international tourism demand can be affected negatively including reduced visitor arrivals, a fall in employment, declines in private sector profits, a reduction in government revenues, and eventually the cessation of further investment" (Huang et al. 2008: 204).

In the case of South Africa, the country as a developing tourism destination has a high rate of crime that negatively affects its tourism competitiveness. The World Economic Forum in its Travel and Tourism Competitiveness Report 2018/19 ranked South Africa at 120 out of 136 countries for perception of safety and security. "Despite the steady increase in popularity with the international community, South Africa has developed a reputation for being an unsafe place to visit. This is not surprising as South Africa has extraordinary high levels of violent crime" (George 2003: 575). The crime situation prevailing in South Africa has given the citizens of the country an undesirable image, as would-be tourists agree on the beauty of the country and comment on their perceptions of safety and security. The City of Cape Town has the highest murder rate in South Africa, and this has led to the deployment of the army (South African National Defence Force) in certain residential areas in Cape Town in July 2019. Sokanyile (2019) noted that there were 3,275 murders reported in the Western Cape in 2018. "In 2018, 83% of all recorded gang-related murders occurred in the Western Cape. Nearly half of all murders were reported to the 26 police stations in gang-infested communities, including areas on the Cape Flats" (Knight 2019: 2). An Anti-Gang Unit was launched in November 2018 to facilitate mitigation of the high rate of gangsterism in Cape Town townships. Van Lennep (2019) noted that the street gangs in Cape Flats have roots in anti-apartheid activism. According to Dziewanski (2015), authorities in the City of Cape Town estimate there are between 100 and 120 gangs in the Western Cape Province, with membership ranging from 80,000 to 100,000.

Annual Komjekejeke traditional festival, Wallmansthal outside Pretoria

In the rural hinterlands outside Pretoria, which is the legislative capital city of South Africa, the AmaNdebele nation undertakes a yearly cultural event to celebrate Ndebele culture and commemorate the life of the late King Silamba. The Komjekejeke Festival, as named, is celebrated at Wallmansthal located close to the N1 Highway North towards the Limpopo Province. The event, launched in 1984 by his descendent and the current King Makhosonke Mabhena II, has grown to become a major traditional ceremony that attracts visitors from all over South Africa. In addition, traditional royalty invited from all over Africa attend this event with their subjects. The event initiated by the Ndebele Royal Family in collaboration with the Ndebele community is an example of a community-initiated cultural celebration that has grown in leaps and bounds, and has remained a niche cultural festival celebrating Ndebele culture.

> The event which celebrates Ndebele culture, arts and heritage seeks also to promote cultural tourism, contribute to economic development and attract investment in the Ndebele Kingdom. Since its inception over three decades ago, the number of tourists attending this event has grown immensely attracting people from South Africa and beyond the borders. It is attended by approximately over 20 000 people from diverse lifestyles. It is also attended by dignitaries including Cabinet Ministers, Members of the Diplomatic Corps, Royalty from different parts of South Africa and beyond the borders including Zambia, Zimbabwe, Botswana, Democratic Republic of Congo to mention a few.
>
> (Sigcau 2017: 1)

Sigcau (2017) noted that the National Department of Arts and Culture (DAC) declared a national heritage site in 1999.

According to Newshub (2016: 1)

> King Silamba was a ruler of the Ndebele nation when they were forcefully moved to KoMjekejeke by the Boers in the late 1860s. Silamba reigned until 1892 and was succeeded by his eldest son Mbhedlengani, whose rule was short-lived... After a bitter conflict over land with Afrikaner farmers, led by JG Bronkhorst, Silamba and some of his followers moved to Komjekejeke in the 1850s. Silamba and four other kings are buried in the vicinity of Komjekejeke. The site was bought by Makhosoke II and is administered under the Silamba Trust.

Dignitaries such as the Premier and high-ranking officials of the Mpumalanga Province, the Mayor of the City of Tshwane, from the Presidency and ministers always attend the yearly commemoration of the Ndebele celebrations at Komjekejeke. As a result, there was a need to upgrade the facilities officially handed over to the AmaNdebele people in 2016. According to the National Department of Tourism (2016) the former Minister of Tourism, Derek Hanekom officially launched and handed over facilities built by the National Department of Tourism (NDT) at the Komjekejeke heritage site, which included a multi-purpose hall and an amphitheatre constructed with the view of driving tourism to the area. The construction process also included new ablution blocks, three new traditional dwellings, 20 picnic sites, renovation of existing buildings including the VIP lapa, existing staff houses and the painting of buildings. In addition, a fully-fledged museum curating the history of the AmaNdebele is located and operational on site. The NDT through its Social Responsibility Implementation (SRI) project funded the entire project at a cost of R22 million. According to Molekwa (2016) the legacy of the investment by NDT includes accommodation that cater to

22 people, a reception area, three administration offices, a restaurant, coffee shop, coffee shop, bathrooms and a trading area.

The NDT promotes the growth of tourism products through infrastructure development under its SRI projects funded by the Expanded Public Works Programme (EPWP). The EPWP involves the creation of labour intensive work opportunities for the unemployed, thereby allowing them to gain employment opportunities that will make them participate in the economy of South Africa. The National Department of Tourism (2019) noted that the criteria for funding SRI projects included that such projects must be located in rural areas, it must unlock tourism potential in the area, the community must benefit directly and the project must improve South Africa's competitiveness as a tourism destination. Another funding criterion was local control. A Silamba Trust was established as the ownership and administrative authority for Komjekejeke. The Silamba Trust has representatives from the Ndebele Royal Family and communities. Molekwa (2016) explained that in the construction stage employment of more than 100 people occurred; at present there are 20 full-time employees to run the site that houses the royal graves, and the Komjekejeke cultural and heritage attraction. In addition, the National Department of Arts and Culture is involved as a funder of the festivities that occur at Komjekejeke. The Department of Arts and Culture (2017) supported the 37th Komjekejeke King Silamba Annual Commemoration through its department's strategic programme of Mzansi Golden Economy (MGE) to reinforce the arts, culture and heritage sector as an economic growth sector, create decent work for local artists and provide skills development for excellence and high performance in the arts culture and heritage sector.

Inferences

The upgrades at Komjekejeke created the conditions to boost tourism by increasing the attractiveness of the culture and heritage products available at Komjekejeke. The development of the museum and Ndebele interpretation centre means product development undertaken to ensure that the Komjekejeke will be able to attract visitors outside the yearly March celebrations as well. Marketing outlays need to be strengthened to attract visitors to come and learn about the Ndebele culture at Komjekejeke. To fully integrate tourism distribution, it is imperative that the attraction at Komjekejeke becomes affiliated as a member of the Tshwane Tourism Association and Gauteng Tourism Authority, so that it can receive more business from industry players. The theoretical framework on Community-Based Tourism (CBT) fits in well at Komjekejeke, which is a community-initiated project that has become a major domestic and regional festival focusing on the celebration of Ndebele culture.

Komjekejeke is located within the municipal borders of the City of Tshwane and identified as an event and festival that can improve the tourism competitiveness of the City of Tshwane. Steyn (2007: 27) it is evident that culture and heritage in a destination play an important role as a key attractor in attracting visitors to a destination. The more attractive a destination is, the more competitive it will be in the longer-term. Some destinations may also capitalise on their cultural and heritage resources to create a sustainable competitive advantage for their destination. However, it is crucial that culture and heritage as a resource is carefully managed for sustainability for enhancing overall destination competitiveness. Finally, the way in which culture and heritage are packaged into accessible, flexible and customised product offerings can also contribute to overall destination competitiveness.

The legacy of hosting the Silamba celebrations at Komjekejeke has been to receive government support that has improved the infrastructure on site that would further enhance on site experience.

Theoretical framework: Community-Based Tourism (annual Komjekejeke traditional festival)

Community Based Tourism (CBT) has often been cited as an alternative to mass tourism and an approach for tourism to become more sustainable. If developed well, CBT can become a poverty alleviation mechanism and a way to access improvements in Quality-of-Life, providing empowerment and greater economic benefit to individuals in local communities.

(Doods et al. 2018: 1548)

CBT has been used to describe a broad range of different tourism models but usually refers to tourism that involves community participation and aims to generate benefits for local communities in the developing world by allowing tourists to visit these communities and learn about their culture and the local environment.

(Lucchetti and Font 2013: 2)

Doods et al. (2018) noted that CBT shares the goals of sustainable development in that it strives to be socially equitable, ecologically sound and economically viable for the long term. The community in CBT is the epicentre of focus and therefore CBT is rooted in community participation, and community development is the desired outcome. Doods et al. (2018) further noted that CBT differs from many other forms of tourism in that it does not solely aim to maximize profits for absent inventors, but to maximize benefits for community shareholders. It is interested in striking a balance between the needs of the community (which is the host community and therefore tourism supply) and the tourists (who are the consumers and tourism demand). "CBT by its very nature is pro-poor, seeking to ensure the majority of the benefits accrue to locals and the local economy" (Strydom et al. 2017: 3).

"CBT is generally small scale and involves interactions between the visitors and host community and is particularly suited to rural and regional areas" (Strydom et al. 2018: 3).

Strydom et al. (2017) noted that characteristics of CBT include local control of development, local participation in decision making and the equitable flow of benefits from the project; the CBT project would incorporate the values and principles of the resident community. This makes it an appropriate vehicle for preserving community heritage. Lucchetti and Font (2013) noted that through local control of tourism businesses and activities, CBT will be able to contribute to cultural and environmental conservation and redistribute the economic benefits amongst the most vulnerable groups such as indigenous communities. CBT is an avenue for diversifying the rural economies to improve the livelihoods, Quality-of-Life and standard of living of indigenous communities (Table 5.1).

> The tourism literature describes CBT in two ways. One of which focuses on community development through tourism, the second on community engagement with the affected community in the planning of tourism as a land use and the subsequent development of a tourism venture to promote long term relationships between service providers and clients.
>
> (Harwood 2010: 1910)

Doods et al. (2018) noted that education and training are key components of capacity building and courses including hospitality and tourism management at the community level. Education and capacity building must improve the knowledge of the community in business enterprises that they are already engaged in, for example, improving the quality of the brick making that the community engages in so that it can lead to more sales orders.

Table 5.1 Key elements for Community-Based Tourism (CBT) success

1.	Participatory planning and capacity building – to strengthen community's tourism management skills
2.	Collaboration and partnerships facilitating links to market – to ensure financial viability
3.	Local management/empowerment of community members
4.	Establishment of environmental/community goals – to ensure outcomes are in alignment with community's values
5.	Assistance from enablers (government, funding institutions and private sector) – to facilitate access to the formal economy
6.	Focus on generating supplemental income for long-term community sustainability

Source: Doods et al. (2018).

CBT should not be viewed as an end in itself, but as a means towards empowering poor communities to take up control over their land and resources, to tap into their potential and to acquire skills necessary for their own development.

(Strydom et al. 2018: 4)

CBT projects can only be successful if they can meet the needs of the market. "Possibly the most important consideration for successful CBT is creating market linkages at the planning stage which can help to develop a market-ready product" (Doods et al. 2018: 1560). To entice the market commercial viability in addition to marketing is important to ensure the success of CBT. "Commercial viability is often enhanced through partnerships practices between CBT enterprises and the private sector that benefit both parties" (Lucchetti and Font 2013: 4).

Recommendations and conclusions

Smaller events have a positive economic impact on the local destinations where they are hosted. In the context of South Africa, in the post-apartheid era, there has been a period of hosting once-off mega-events that have limited and debatable impact on the economic fortunes of South Africa. The legacy of the 2010 FIFA World Cup in terms of stadia is not debatable, but the cost of hosting the mega-event remained high, using long-term debt to fund the mega-event. The era of hosting mega-events in South Africa is over due to low government finances. Small-scale events that do not depress state finances are much more sustainable. Event legacies for small-scale events are much more profound as they have an immediate positive impact on the host community. Ramadan became a lobbying point to acquire heritage protection for the Bo Kaap. The high rate of crime in Cape Town ceases during Ramadan. Komjekejeke heritage declaration occurred in 1999 and in 2016; the infrastructure was improved and handed over to the Ndebele nation. This is an example of successful CBT, as the event is community initiated.

There is scope for improving the events much further. Security is now a major issue in event management, and principles of event management are implemented from inception to close down of the events. The professionalisation of the above-mentioned events is imperative as a means of building lasting legacies as part of ensuring the long-term sustainability of these events. Evaluating the impact of the events is important to investigate strategies for improving these recurring events. Patron satisfaction is a major issue requiring investigation, in line with the principles of continued cycle of improvement. This would ensure that needs of current and potential patrons are understood by event organizers as part of sustained intelligence. These measures will ensure that the events become more attractive in attracting public and private funding, and sponsorship that could improve the financial standing of these events, especially in ensuring there is a Return-on-Investment (ROI).

References

Booysens, I. & Visser, G. (2010). Tourism SMME development on the urban fringe: The case of Parys, South Africa. *Urban Forum*, 21: 367–385.

Botha, A. (2012). Modelling Tourism Demand for South Africa Using a System of Equations Approach: The Almost Ideal Demand System. Trade & Industry Policy Strategies. Retrieved from: https://www.tips.org.za/index.php [Accessed 29 October 2012].

Chok, S., Macbeth, J. & Warren, C. (2007). Tourism as a tool for poverty alleviation: A critical analysis of 'pro-poor tourism' and implications for sustainability. *Current Issues in Tourism*, 10 (2 & 3): 144–165.

Civil Defence Emergency Management Act (2002). http://www.legislation.govt.nz/act/public/2002/0033/51.0/DLM149789.html

Cornelissen, S. (2010). The geopolitics of global aspiration: Sport mega-events and emerging powers. *The International Journal of the History of Sport*, 27 (16–18): 3008–3025.

Dangor, S. (1997). The expression of Islam in South Africa. *Journal of Muslim Minority Affairs*, 17 (1): 141–143.

David, N.M. (2015). *Islamic Tourism in South Africa: An Emerging Market Approach*. Hesley: IGI Publishing.

Department of Arts and Culture. (2017). Deputy Minister Mabudafhasi to Address the 37th komjekejeke King Silamba Annual Commemoration. Retrieved from: http://www.dac.gov.za/content/deputy-minister-mabudafhasi-address-37th-komjekejeke-king-silamba-annual-commemoration-0 [Accessed 1 July 2019].

Doods, R., Ali, A. & Galaski, K. (2018). Mobilising knowledge: Determining key elements for success and pitfalls in developing community based tourism. *Current Issues in Tourism*, 21 (13): 1547–1568.

Dziewanski, D. (2015). South Africa: Escaping Gangland through Islam. Retrieved from: https://www.aljazeera.com/indepth/features/2015/06/south-africa-escaping-gangland-islam-150609081659960.html [Accessed July 2019].

Ferreira, S. (2007). Role of tourism and place identity in the development of small towns in the Western Cape, South Africa. *Urban Forum*, 19: 191–209.

Ferreira, S.L.A. (1999). Crime: A threat to tourism in South Africa. *Tourism Geographies: An International Journal of Tourism Spaces, Place and Environment*, 1 (3): 313–324.

Fourie, J., Rosello, J. & Santana-Gallego, M. (2014). Religion, religious diversity and tourism. Stellenbosch Economic Working Papers: 09/14. Stellenbosch: Bureau for Economic Research.

George, R. (2014). *Marketing Tourism in South Africa*. Fifth Edition. Cape Town: Oxford University Press.

George, R. (2003). Tourist's perception of safety and security while visiting Cape Town. *Tourism Management*, 24: 575–585.

Hajubaba, H. & Dolnicar, S. (2017). How to prevent tourists from canceling when a disaster hits the destination: Promising measures, crisis-resistant target segments and leveraging peer-to-peer networks. *Travel and Tourism Research Association: Advancing Tourism Research Globally*, 22: 1–7.

Hajubaba, H., Gretzel, U., Leish, F. & Dolnicar, S. (2015). Crisis-resistant tourists. *Annals of Tourism Research*, 53: 46–60.

Hamza, I.M., Chouhoud, R. & Tantawi, P. (2012). Islam tourism: Exploring perceptions & possibilities in Egypt. *African Journal of Business and Economic Research*, 7 (1): 85–98.

Harwood, S. (2010). Planning for community based tourism in a remote location. *Sustainability*, 2: 1909–1923.

Haupt, T. (2017). Halal tourism getting stronger than ever. *Tourism Review*. Retrieved from: https://www.tourism-review.com/halal-tourism-growing-in-popularity-news [Accessed July 2019].

Henama, U.S. (2017). Marikana: Opportunities for heritage tourism. *African Journal of Hospitality, Tourism and Leisure*, 6 (4): 1–16.

Henama, U.S. (2016a). Nkandla: The unexplored frontier of heritage tourism in Zululand, South Africa. *African Journal of Hospitality, Tourism and Leisure*, 5 (2): 1–11.

Henama, U.S. (2016b). The low cost carrier bandwagon: Lessons for Skywise Airline. *African Journal of Hospitality, Tourism and Leisure*, 5 (2): 1–23.

Henama, U.S. (2014). The demise of 1Time Airline and the reaction of various interest groups. *African Journal of Hospitality, Tourism and Leisure*, 3 (2): 1–11.

Huang, Y.C., Tseng, Y.P. & Petrick, J.F. (2008). Crisis management planning to restore tourism after disaster. *Journal of Travel & Tourism Marketing*, 23: 203–221.

Kamish, R. (2019). Beautiful Cape Malay traditions you'll notice during Ramadaan. *Cape Town Magazine*. Retrieved from: https://www.capetownmagazine.com/ramadaan [Accessed July 2019].

Knight, T. (2019). Hanover Park residents plan shutdown to highlight gang violence. *Daily Maverick*. Retrieved from: https://www.dailymaverick.co.za/ article/ [Accessed January 2019].

Labuschangne, V. & Saayman, M. (2014). Role of location in the attendance and spending of Festinos. *African Journal of Hospitality, Tourism and Leisure*, 3 (1): 1–11.

Lubbe, G. (1987). Robben Island: The early years of Muslim resistance. *Kronos: South African Histories*, 12: 49–56.

Lucchetti, V.G. & Font, X. (2013). Community based tourism: Critical success factors. ICRT Occasional Paper No. 27. Retrieved from: https://www.icrtourism.org [Accessed January 2019].

McDonald, L., & Small, J. (2017, August 23). Christchurch convention centre cost updated to $475m. *The Press*. Retrieved from https://www.stuff.co.nz/the-press/news/96090557/convention-centre-cost-updated-to-475m

Molekwa, S. (2016). Tourism Continues to Transform Lives through SRI Bojanala. Retrieved from: https://www.tourism.gov.za/AboutNDT/Publications/Bojanala%20January-March%202015.pdf/ [Accessed January 2019].

Namin, T.A.A. (2013). Value creation in tourism: An Islamic approach. *International Research Journal of Applied and Basic Services*, 4: 1253–1264.

National Department of Tourism. (2019). The National Department Of Tourism's (NDT) Social Responsibility Implementation (SRI) Project. Retrieved from: https://tkp.tourism.gov.za/lg/support/pages/current-sri-projects.aspx/ [Accessed January 2019].

National Department of Tourism. (2016). Minister Hanekom to launch tourism facilities at Komjekejeke and engage with the local community. Retrieved from: https://www.tourism.gov.za/ [Accessed January 2019].

Newshub. (2016). AmaNdebele Nation to Commemorate Late Ndebele King Silamba. Retrieved from: https://za.newshub.org/amandebele-nation-commemorate-late -ndebele-king-silamba-20948779.html#/ [Accessed January 2019].

North West University. (2017). South Africa Must Reap the Benefits of Halal Tourism. Retrieved from: http://www.nwu.ac.za/content/ [Accessed January 2019].

Onishi, N. (2016). Muslim enclave forged in apartheid now faces gentrification. *The New York Times*. Retrieved from: https://www.nytimes.com/ [Accessed January 2019].

Osland, G.E., Mackoy, R. & McCormicj, M. (2017). Perceptions of personal risks in tourists' destination choices: Nature tours in Mexico. *EJTHR*, 8 (1): 38–50.

Pather, R. (2018). Throwing the boeka at crime: How Muslims are reclaiming the streets. *Mail & Guardian*. Retrieved from: https://mg.co.za/article [Accessed January 2019].

Rogerson, C.M. & Rogerson, J.M. (2014). Agritourism and local economic development in South Africa. *Bulletin of Geography, Socio-economic*, 26: 93–106.

Sigcau, S. (2017). King Silamba commemoration on 4 March 2017. *The Diplomatic Society*. Retrieved from: http://www.thediplomaticsociety.co.za/archive/archive/ 2072-king-silamba-commemoration-on-4-march-2017/ [Accessed July 2019].

Sokanyile, A. (2019). Call for Anti Gang Unit to spread out across Cape Town as violence escalates. *IOL*. Retrieved from: https://www.iol.co.za/weekend-argus/news/ [Accessed January 2019].

South African Historical Organisation. (2019). *Malay Quarter, Cape Town*. Retrieved from: http://www.sahistory.org.za/places/malay-quarter-cape-town/ [Accessed January 2019].

Spencley, A. & Meyer, D. (2012). Tourism and poverty reduction: Theory and practice in less economically developed countries. *Journal of Sustainable Tourism*, 20 (3): 297–317.

Steyn, T. (2007). The Strategic Role of Cultural and Heritage Tourism in the Context of a Mega-Event: The Case of the 2010 Soccer World Cup. Published Masters-Thesis. Pretoria: University of Pretoria.

Steyn, J.N. & Spencer, J.P. (2011). South African tourism: An historic evaluation of macro tourism policies. *African Journal of Physical, Health Education, Recreation and Dance*, 17 (2): 178–200.

Strydom, A.J., Mangope, D. & Henama, U.S. (2018). Economic sustainability guidelines for a community-based tourism project: The case of Thabo Mofutsanyane, free state province. *African Journal of Hospitality, Tourism and Leisure*, 7 (5): 1–17.

Strydom, A.J., Mangope, D. & Henama, U.S. (2017). Economic sustainability guidelines for a community-based tourism project: The case of Thabo Mofutsanyane, free state province. *African Journal of Hospitality, Tourism and Leisure*, 6 (1): 1–17.

Strydom, A.J., Saayman, A. & Saayman, M. (2006). The economic impact of the Volksblad Arts Festival. *Acta Commercii*, 87–98.Torres, R. (2003). Linkages between tourism and agriculture in Mexico. *Annals of Tourism Research*, 30 (3): 546–566.

TNT News. (2017). Cape Town seeks to become Halal tourism destination. Retrieved from: http://tntnews.co.za [Accessed January 2019].

Umar, U.M. (2017). Halaal tourism industry in South Africa under development. *Radio Islam*. Retrieved from: http://www.radioislam.org.za/index.php/latest-news/ [Accessed January 2019].

United Nations Conference on Trade and Development. (2010). The contribution of tourism to trade and development. Second Session, Geneva, 3–7 May 2010, Item 5 of the TD/B/C.I/8, United Nations Conference on Trade and Development, Geneva, Switzerland.

Van Lennep, T. (2019). Cape Town Gangs: Political Dimensions. Helen Suzman Foundation. Retrieved from: https://hsf.org.za/publications/hsf-briefs/cape-town-gangs-political-dimensions. [Accessed January 2019].

World Economic Forum. (2018). Travel and Tourism Competitiveness Report 2018/19. Geneva: WEF.

Zamani-Farahani, H. & Eid, R. (2016). Muslim world: A study of tourism & pilgrimage among OIC Member Countries. *Tourism Management Perspective*, 19: 144–149.

6 Entrepreneurship development and Slow Food events

Burcin Kalabay Hatipoglu, Onno Anıl,
Saadet Memiş and Dilan Şahin

Introduction

The present study was conducted in the rural setting of Şile, a Black Sea coastal town 60 kilometers northeast of the Istanbul city center. The Şile Earth Market, a Slow Food network activity, is the third one in the country (Hatipoglu, Aktan, Duzel, Kocabas, and Sen, 2016). The market aims to create income for local farmers and producers while providing healthy and nutritious food to visitors to the market. Residents of Şile are frequent visitors to the biweekly market, which also attracts visitors from nearby regions and food lovers from Istanbul. Although rural events tend to have low demand and low value (Getz, 2008, p. 407), the Şile Earth Market attracts thousands with their food-themed events (e.g. Terra Madre Day and Seed Exchange Day). At the same time, the market space and the surrounding streets have become a social and cultural venue for the community. Hence, the Earth Market is not only a farmer's market but also a planned tourism event whose goals include adding value to the livelihood of the community and bringing food-based tourism to the region.

This chapter examines how participants of the market and tourism businesses are empowered in the decision-making and implementation processes of the Şile Earth Market. Observations and interviews with farmers and producers, tourism businesses, non-profit organizations, and the municipality formed the data set. It is essential to uncover how businesses involved in the market are empowered in order for researchers to make recommendations for improvements in tourism planning. Interest in the impact of tourism events is relatively recent (Getz, 2008), allowing for exploration before agreeing on an integrated measure that uses sustainability dimensions (Getz and Page, 2016a, p. 41). The chapter suggests a framework for supporting rural entrepreneurship through a planned tourism event while adopting a sustainable tourism development lens. The chapter will be of interest to public officials, local organizations, and tourism professionals.

Background of the study

Event tourism and sustainable tourism development

Event tourism is a subfield of tourism management characterized as being "inclusive of all planned events in an integrated approach to development and marketing" (Getz, 2008, p. 405). As widely recognized, planners view tourism events as a tool for promoting regional development. Tourism events attract tourists in off-peak seasons, help increase infrastructure and destination capacity, improve the destination image, contribute to place marketing, and emphasize the unique feature of the destination (Getz, 2008, p. 405). Besides contributing to the local economy, festivals and events build social capital, enhance community capacity, and support non-tourism-related products and services.

Tourism activities may create negative externalities for a destination when not well planned. Measuring the impact of events will be of importance for various reasons such as communicating to the stakeholders and assuring transparency, receiving funding, place, and permits, and maintaining legitimacy. Scholars suggested many models for measuring the goals and impacts of festivals and events, most of which concentrate on economic outputs. Using a more holistic approach, Sherwood (2007) has developed environmental, economic, and social impact measurements for tourism event assessment. Getz (2008) proposed to add personal and cultural outcomes to Sherwood's (2007) approach in evaluating event outcomes.

A sustainable tourism development approach makes optimal use of assets while minimizing costs. Furthermore, it recommends that tourism be developed considering "current and future economic, social and environmental impacts, addressing the needs of visitors, the industry, the environment, and host communities" (United Nations Environment Programme; World Tourism Organisation, 2005, pp. 11–12). Thus, in applying sustainable tourism development in rural areas, the informed participation of all relevant stakeholders and consensus-building are critical factors. Community-run events involve a complex set of stakeholders, including facilitators, co-producers, suppliers, the audience, and the impacted community (Capriello and Rotherham, 2013). Building partnerships with stakeholders in tourism planning and management has become a common way of engaging stakeholders. In event tourism, involving the host community, generating trust, and maintaining commitment among local suppliers are required for achieving socio-economic objectives (Capriello and Rotherham, 2013). Government agencies play a strategic role in the creation and development of healthy democratic practices (Dredge and Whitford, 2011) for long-term value creation.

Acknowledging the complexities of planned event tourism, Getz and Page (2016b) offer a framework for studying event tourism. The framework places event tourism experiences in the middle of the model, and it constructs a circle where personal antecedents and decision-making and planning, design, and management (i.e. stakeholders, organizations) build the experience. Through various processes (spatial and temporal, policy and knowledge creation), outcomes and impacts (*personal, societal, political, cultural, economic, and environmental*) are produced. In reviewing the framework of Getz and Page (2016b), Higgins-Desbiolles (2018) identifies the *community* as the missing pillar of tourism event management. She challenges the limited understanding of community participation and builds a case around communities getting actively involved in the decision-making processes of sustainable event tourism.

Rural entrepreneurship

Entrepreneurship is often associated with competitiveness, development, and economic growth. For these reasons, increasing entrepreneurial capacity in both urban and rural areas is desirable. In rural regions, small businesses are the backbone of the economy. Entrepreneurial activities include a broad spectrum, at one end of which there are fast-growing small businesses that use new technologies. On the other end, there are slow-growth, more traditional family owned businesses. The *World Encyclopedia of Entrepreneurship* (McElwee and Atherton, 2011) defines a rural entrepreneur as:

> An individual who uses the resources of the regional economy.... In order to gain competitive advantage by trading goods or services, which ultimately generate social or economic capital for the rural environment in which the entrepreneur is located.
>
> (p. 382)

Embeddedness in the local resources is thus an inseparable component of rural entrepreneurship. Moreover, a supportive environment is a condition for entrepreneurial success. The local governmental agencies and other local entities can play strategic roles in facilitating entrepreneurs' access to networks and critical resources.

Starting and growing a new business anywhere is challenging. In addition to the well-documented obstacles of developing a new business – financing, marketing, and management—rural areas have additional challenges of low population density and remoteness. Low local demand prevents businesses from achieving economies of scale and growth. Rural areas have limited access to services readily found in urban areas such as banks, credit institutions, and business support systems. Access to labor markets and distribution channels is also problematic (McElwee and Atherton, 2011). In many cases, talented young people abandon rural areas for better jobs in cities, leaving behind older people and the less talented. The type of businesses that can start and prosper in rural areas is affected by remoteness from urban areas, highways, and airports.

Recently, people living in urban regions have become attracted to rural regions for their quality of life. Tired of the fast pace and sameness in busy cities, city dwellers are interested in rural regions with distinct local characteristics. This interest has created an opportunity for rural entrepreneurs to generate new income sources from agricultural and non-agricultural activities. Tourism that is developed without considering the needs of a community can do more harm than good to a destination. However, tourism that is responsibly planned can improve the life quality of the residents. Tourism entrepreneurship has the potential to stop outward migration from rural areas while strengthening the local culture and identity (Lordkipanidze, Brezet, and Backman, 2005).

Empowerment for sustainable tourism development

Local attitudes toward tourism development carry importance, as they determine residents' willingness to engage in tourism activities. Social exchange theory explains this relationship as the residents evaluating the perceived benefits and costs before giving support to new tourism development. Despite the importance of the local population in tourism development, some scholars have argued that the stakeholders also need to be empowered for tourism to be sustainable. Scholars have shown that tourism empowers community members socially and psychologically (Butler, 2017; Moswete and Lacey, 2015).

Scheyvens (1999) identifies four components of empowerment. Economic empowerment occurs when economic gains result from formal and informal sector employment, there are new business opportunities, the income is equitably distributed in the community, and the results are long-lasting. Social empowerment happens when community cohesion and collaboration are improved based on tourism activity (Scheyvens, 1999, p. 248). Psychological empowerment relates to the pride that residents have in their traditions and culture, strong confidence in themselves and who are self-dependent and optimistic about their future (Scheyvens, 1999, p. 248). When a community is politically empowered, a wide range of individuals (e.g. youth and women) take part in the decision-making processes (Scheyvens, 1999, p. 248).

Accepting the importance of empowerment in sustainable tourism development, Boley and McGehee (2014) developed the Resident Empowerment through Tourism Scale (RETS), which operationalized the concept. Using the RETS scale, Boley, McGehee, Perdue, and Long (2014) examine residents' perceptions of the positive and negative impacts of tourism in the US state of Virginia. The results highlighted the role of both economic and non-economic factors in resident support for tourism. Using the same model, a qualitative study conducted by Boley and Gaither (2016) revealed both the positive and negatives aspects of cultural heritage tourism development.

The Slow Food movement and the Earth Market

The Slow Food movement started in Italy as a protest against fast-food chains and fast living. Slow Food organization aims to protect local food cultures and traditions and to interest people in what they eat. Since the organization's official foundation in 1989, interest in the movement has grown, as there are now more than one million supporters and one hundred thousand members of the organization in 160 countries (Slow Food, 2020). The Slow Food movement has been influential in advancing slowness in tourism.

Ark of Taste, Slow Fish, and Earth Market are just a few of the projects that help the Slow Food organization disseminate its philosophy of good, clean, and fair food. The 69 Earth Markets around the world bring together small-scale farmers and consumers to promote local and seasonal food (Slow Food, 2020). In these markets, farmers take responsibility for their products, and they find opportunities to start dialogues with consumers. Previous research suggests that consumers who shop at an Earth Market find the produce tasty, high quality, fresh, local, seasonal, and safe (Bazzani, Asioli, Canavari, and Gozzoli, 2016). Earth Markets differ from traditional farmer's markets in that farmers must produce the food (e.g. fruits, milk, vegetables) locally and sustainably. Farmers should attend the market themselves and sell products directly to the consumers at a price that is fair for both parties. Taste education, described as "the reawakening and training of the senses and the study of all aspects of food and its production," is a crucial component of the Earth Market (Slow Food, 2020).

The Earth Market in Şile, a small town near Istanbul, started with the leadership of the Slow Food Şile-Palamut Convivium in 2015. Two other local non-profit organizations (the Şile Tourism Association and the Ovacık Village Women's Seed Association), with the cooperation of the Istanbul and Şile municipalities, supported the formation and development of the market. The market is open on Fridays and Sundays in the downtown area of Şile, with approximately 60 producers. Farmers sell local produce (mostly greens, eggs, and milk) along with secondary products (e.g. bread, vinegar, and pastries). The farmers come from the various villages around Şile, and there is a waiting line for joining the market. The organizing committee of the market chooses the producers from a waiting list every six months; those who do not obey the market rules are banned from participation. The Şile municipality controls the production and sale of the goods. The Slow Food Şile-Palamut Convivium consistently provides support to the farmers by educating them and connecting them with a broader network.

Study setting

Şile, a coastal town on the Black Sea, is one of the larger boroughs of Istanbul. Its remoteness from the city center and its vast forests have kept it relatively less occupied than other boroughs of the city. The population is a little over 35,000, but during the summer months, the population exceeds a million. The town has 58 villages that people from different ethnic heritages occupy. Tourism, forestry, agriculture, and fishing are the primary sources of livelihood in Şile. The beaches attract hordes of visitors during the summer months, which pushes the infrastructure to its limits. In the last few years, the municipality has tried to change the branding of the town to turn it into a food and nature-based tourism destination. This change should eliminate seasonality and increase the economic value of tourism without exerting more pressure on the life quality of the residents.

Research methods

To investigate a food-based event's impact on entrepreneurship development in rural areas, we applied a qualitative approach in a case study of the Şile Earth Market. Given the aim of our study and the chosen research method, the event described had to be local, planned, periodic, and focused on the improvement of rural entrepreneurship. The Şile Earth Market met all the defined criteria.

Research design

We organized the data collection in Şile in several stages between December 2015 and October 2017. The first stage relied on secondary sources for investigating the central themes of the study. Furthermore, these sources aided to advance a contextually appropriate research design. In the second stage, we interviewed the representatives of the founding non-profits and the local municipality. In March 2016, we pilot-tested the interview guideline and duly revised it. We conducted the interviews with stakeholders during April and May 2016. The last stage of data collection involved follow-up interviews with the respondents between September 2016 and October 2017.

Data collection and the survey

We employed participative observation, surveys, and semi-structured interviews for primary data collection. The structured participative observations aimed to complement the findings of the surveys. During the study, we attended numerous events organized at the Earth Market. We documented the interactions between farmers and consumers by taking photographs and detailed notes. The empowerment framework of Scheyvens (1999) guided the definition of variables: economic, psychological, social, and political empowerment. In addition to six demographic questions, the interview script to be used with the stakeholders included four questions on tourism development in Şile and 15 questions on empowerment that we operationalized in line with the RETS scale of Boley et al. (2014). We conducted the interviews on market days, and each lasted between 15 minutes and 60 minutes. We obtained consent for recording and transcription. We organized interviews with the Slow Food Şile-Palamut Convivium (non-profit 1), the Şile Tourism Association (non-profit 2), the Ovacık Village Women's Seed Association (non-profit 3), and the Municipality around predetermined themes. These included the history of the market, partnership building and planning, resources shared with farmers and producers, issues in the implementation stage, and plans.

Sample

We employed a purposive sampling approach, and we chose tourism stakeholders in downtown Şile as our sampling universe. It included the 57 farmers and producers that take part in the Earth Market and tourism businesses that are potentially affected by the Earth Market event (40 restaurants, 32 accommodation facilities, eight transportation companies, 32 souvenir shops). We also conducted interviews with the founders and the supporters of the Earth Market (three non-profits and the Şile Municipality).

Analysis and construction of the framework

First, we coded the data collected from the participating farmers and producers, and tourism businesses using the predetermined categories of the empowerment framework. We employed content analysis to define commonalities and differences. Second, we thematically analyzed the data obtained through observations and interviews with non-profits and the Şile Municipality. At this stage, the framework for studying event tourism, as suggested by Getz and Page (2016b), guided the coding process. In analyzing the content of the themes, we paid attention to the emerging patterns, and we recorded any new themes that might add insights to the understanding of the Earth Market event. We classified the findings according to the categories of event drivers, actors of planning and implementation, resources shared, the event, outcomes, and impacts. The structuring of the findings resulted in a framework for supporting rural entrepreneurship through a planned event (Figure 6.1).

Findings

We conducted interviews with a total of 82 stakeholders (48.5% of the sample universe), 55 of which were tourism businesses (24 restaurants, four accommodation facilities, two transportation companies, 11 souvenir shops, and four other). Of the 57 farmers and producers of the Earth Market at the time, the researchers were able to reach 37. Some come to the market only during the summer months. The demographic breakdown of all respondents in terms of biological sex were 29 females (35.4%) and 53 males (64.6%). The ages of the participants were distributed relatively evenly, 18 between the ages of 15 and

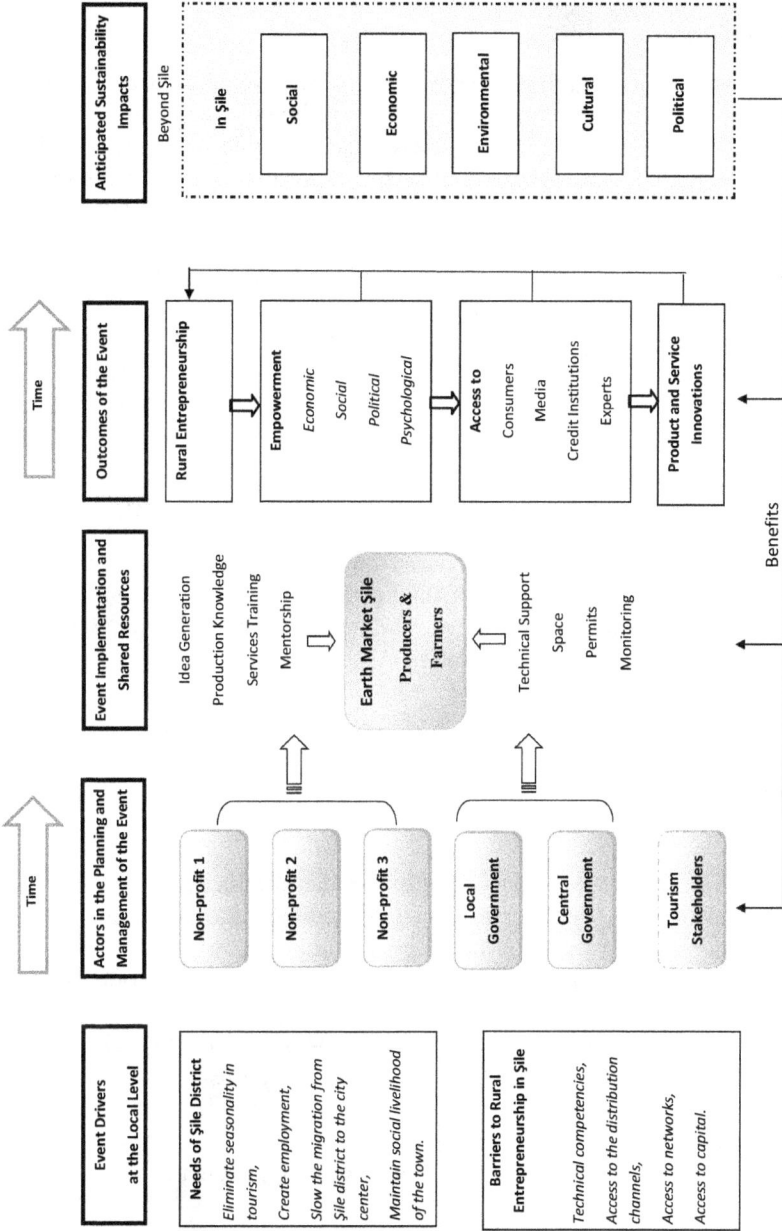

Figure 6.1 A framework for supporting rural entrepreneurship through a planned event.

25, 21 between the ages of 26 and 40, 23 between the ages of 41 and 55, and 20 aged 55 and over. Less than half of the participants (45.1%) had primary school education, and the rest had a high school diploma or higher (54.9%).

Empowerment

To explore to what degree the tourism event empowered the tourism stake-holders, we asked questions regarding all four dimensions. Their answers indicate that all groups perceive themselves as having been empowered, least economically and most psychologically. In other words, the results suggest that stakeholders are proud of the touristic attractiveness of Şile as a destination, and they like sharing their feelings with visitors. A young participant of the Earth Market expressed her *psychological empowerment* as follows:

> My mother makes delicious bread and food items. We sell these here every week at the market. I enjoy it when customers taste her delicious and healthy food.
>
> (Participant 4, Female)

Participants do not feel they have been significantly empowered economically from the event, but they are confident that the situation will get better in the future. They state that income from the market complements their family income. One participant of the Earth Market explains how this new income has lessened her insecurity and has given her the chance to provide her children with better opportunities:

> I'm happy to earn extra money because I can give an allowance to my children, and I also want to be able to support my children's education in the future without depending on others.
>
> (Participant 3, Female)

We find that various categories of empowerment influence each other. For instance, as the Earth Market farmers and producers started gaining an economic outcome, the respect of their family members also improved (political empowerment). Non-profit 3 described the transformation as follows:

> The daughter of one of the women farmers was hesitant to visit her mother at the market after school. She did not see it as a prestigious job, and she did not want any of her friends to know that her mother had a stall in the market. However, as time passed and her mother started earning money, we started seeing her here every market day, helping her mother.

Some of the female farmers and producers, who were initially quite shy, started displaying entrepreneurial characteristics such as ambition and proactive thinking. They also show an understanding of market demand.

Participant 25 acknowledged that in order to generate more income from the market, she needed to expand her product variety. Another participant shared her plans for expanding her business:

> Some of the women that started earlier than us in the market got bank loans and built their kitchen for bread-making. They paid off their loans and even bought a condo for their children. I am considering doing the same eventually.
>
> (Participant 7, Female)

On the other hand, tourism businesses have not yet realized the potential economic spillover benefits of the Earth Market as a tourism event. Most of them are doubtful of its economic benefits for the town. Only some of the café owners acknowledge that visitors to the market also stop at their businesses for food.

The farmers and producers perceived to be *socially* empowered due to joining the market. They admit that they have no time for anything except their farming and production activities and days spent at the market, but after joining the market, they made new friends from other villages. We observed social cohesiveness at informal gatherings such as birthday celebrations at the market. One of the participants described her new social life:

> I have come together with other villagers. They had seeds I did not have, so we exchanged seeds. Now we spend time together at the market, and we even have our social gatherings here.
>
> (Participant 16, Female)

The founders of the event (the non-profits and the Şile Municipality) closely observed the social transformation of the farmers and producers in the market. Non-profit 2 notes his observations:

> When the market first started, the farmers and producers would not make eye contact with visitors or engage them in a dialogue. Over time, we managed to convince them of the benefits of engaging with their customers and helped them understand the mission of Slow Food for food tasting. Now they all have food samples in front of their stalls, and they encourage their customers to try new recipes using the fresh produce they purchase.

Confirming the above statement, we also witnessed that the farmers and producers made friends with visitors, even sharing updates on social media. They became socially active thanks to the market which allowed them to spend time outside of their villages. Further, some of them perceived to be socially empowered even if they did not feel to be economically empowered.

Farmers and producers also perceived that they had become empowered *politically*, that is, they now have a role in making decisions about the Earth

Market. One participant pointed out that they now have more access to the Şile Municipality and that the Municipality has started to communicate with them directly (Participant 2). Another declared that a family member started to take her more seriously and asked for her opinions in decision-making more often (Participant 3). Non-profit 1 and non-profit 2 confirmed the same and described how some of the women became more persistent in their attempts to influence the governing committee.

In order to identify differences between family members in terms of perceived empowerment resulting from the Earth Market, we compared the answers of female and male Earth Market participants. The expectation was that the female participants would feel more empowered than male participants, but their responses revealed just the opposite. Our observations also confirmed this finding, showing that family members worked together throughout the production and sales processes. Many of the women participants said that their husbands were supportive of their entrepreneurship activities. One male participant shared his view of co-entrepreneurship as follows:

> Without the support of family members, we would not be able to come to the market. This market encourages us to work together. My wife has changed positively, and she has made new friends at the market.
>
> (Participant 25, Male)

We asked four open-ended questions in the survey in order to depict how tourism stakeholders view tourism in Şile and to identify their expectations. The results were diverse, as there were both positive and negative perceptions of tourism. The stakeholders believe that visitors enjoy Şile's sea, nature, history, and farmlands. Their expectation from tourism is that it should contribute to the local economy, and they view proximity to Istanbul as an advantage for tourism development. However, they are also aware that tourism creates specific problems for the town, and they expect the Şile Municipality to find solutions. One of the many concerns is the lack of entertainment activities in the town, for residents and visitors alike.

Supporting rural entrepreneurship

In this section, we present the results as the drivers of the event at the local level (needs of the local community and barriers to entrepreneurship), actors involved in the planning and managing of the event, implementation of the event, outcomes, and anticipated sustainability impacts. We display these elements of the event in Figure 6.1.

Needs of the local community: The Şile Municipality, taking into consideration the low income-generating potential of summer tourism in Şile and the stress it creates on the infrastructure, considers developing alternative tourism types (e.g. culture, food, and nature-based tourism). It plans to do so by building upon the town's assets, which include the traditional Şile woven cloth, agriculture, and food-based activities. The Şile municipality also searches for ways to create alternative jobs to forestry (the villagers are now restricted to the amount of wood they can cut from the forests) and to slow the migration to Istanbul. Besides, the community expects the Municipality to maintain the social livelihood of the town by planning new events (e.g. seasonal music festivals) or supporting new entertainment areas.

Planning phase: The Şile Earth Market started with the leadership of Slow Food Şile-Palamut Convivium (non-profit 1), the contribution of the Şile Tourism Association (non-profit 2), and the Ovacık Village Women's Seed Association (non-profit 3). Each non-profit encouraged villagers to take part in the market. A producer shares her experience of the first days:

> We were only five at first. They (non-profit 1 and non-profit 2) convinced us to take part, but whatever they said the rules were, we accepted all of them. We have come a long way, look how crowded the market is now!
> (Participant 6, Female)

The non-profits also convinced the Şile Municipality to issue permits, provide a permanent market space, and help organize and promote the event. Non-profit 2 visited similar markets around the country and shared the findings with the municipality regarding the proposed location of the market, the physical needs of the market (a semi-covered space, a parking lot, toilets), and the most suitable days for opening the market. At the request of the local government, the Istanbul Municipality built the infrastructure. During the planning and implementation stages, the partnership between the three non-profits and the Şile Municipality encouraged the farmers and producers to engage in activities and start their family businesses.

Event implementation: The literature records multiple obstacles to rural entrepreneurship in the literature, but in Şile, rural entrepreneurs have overcome some of these barriers because of the *resources shared* and *support* provided by the founders. The non-profits have been influential in idea generation, product support (e.g. Slow Food principles, the use of heirloom seeds, the planting and selling of local in-season products, the protection of local recipes), mentorship, and services training (e.g. fair prices for goods, clean and fair food, educating customers on local food, dialogue with visitors). For two consecutive years, the non-profits financed the travel and accommodation for two young farmers to join them at the Slow Food festival in Torino, Italy. Due to these educational activities, the technical and interpersonal skills of the entrepreneurs improved, and they gained self-confidence. The local government provides the space, permits, and technical support for the farmers and continues to monitor the functioning of the market. The Şile Municipality employs two experts to visit farms regularly.

The local branch of the Turkish Ministry of Agriculture ensures minimum use of pesticides, closely monitors production processes, and recommends alternative methods for sustainable production when needed.

Despite their differences, the partnership model built between the three civil society organizations and public entities is exemplary for events management and sustainable tourism development. In order to reach a common goal, partners of the Earth Market achieved high levels of resource sharing and high interdependence amongst themselves (Austin and Seitanidi, 2012). They established a committee that meets regularly to solve problems during implementation and to oversee plans and activities. Furthermore, partners that have rallied around a shared vision have put aside their agendas and are working collaboratively in order to provide value to the town. The committee enables stakeholder involvement by including representatives from the non-profits, the municipality, tourism businesses, and the farmers and producers.

Outcomes of the event: In evaluating the *outcomes* of the market, it is possible to note the effects on rural entrepreneurship and product and service innovations. All the farmers and producers registered themselves on the official farmer registration system and as a result they can receive financial support from the government. The farmers and producers who become psychologically, socially, politically, and economically (in order of importance) empowered have overcome some of the barriers to rural entrepreneurship with the support that they receive to continue growing their businesses. They now have access to credit institutions, media, experts, and customers. For example, national and local television programs covered Şile and the Earth Market. Local and international chefs are frequent visitors to the market, all of whom shared their visits on social media. The Earth Market hosted special themed events, including an international honey festival and an event where experts shared knowledge on the use of local medicinal herbs. Non-profits have been mentoring the farmers on packaging, branding, and ways to bring forward the value of their products. These mentoring activities have resulted in new product development (e.g. honey vinegar, tarhana soup with nettle) and service innovations (serving breakfast to groups during special events).

The empowerment of the participants, together with the fact that they were able to overcome some of the barriers to rural entrepreneurship, has started a *positive loop* in the market (Figure 6.1). In this new *loop,* empowered entrepreneurs that are interested in growth are engaging in product and service innovations and reinvesting in their businesses. In the case of the Earth Market, in order to expand their businesses, rural entrepreneurs involved family members and neighbors in production and service, secured bank loans for reinvestment, received additional training, and some have gone on to open shops in downtown Şile. The success of the Market influenced some of the residents who have not been previously involved in farming or production and they also started taking part in entrepreneurial activities.

Anticipated sustainability impacts: In addition to rural entrepreneurship outcomes, we anticipate the market to have economic, psychological, environmental, social, cultural, and political sustainability impacts in the medium and long term – and beyond the destination. The event was instrumental in assisting the protection of the local heirloom seeds. The recipes preserved through the market will help maintain the food-related heritage of the town. Using local products in season and educating the young generation on good food will encourage healthy living. The use of heirloom seeds and restrictions on the use of pesticides support biodiversity, so we expect these to have long-term positive environmental effects. Moreover, women finding alternative income opportunities and receiving an education while balancing their family and social lives will set an excellent example for the next generations.

In addition to sustainability impacts, we expect the event to have *benefits* for tourism stakeholders as well as for the partners of the event. The fact that the event ensured the engagement of locals and displayed a multistakeholder partnership model makes this case a valuable example for practitioners and researchers. With this in mind, researchers and representatives of municipalities in other parts of Turkey frequently visit the Şile Earth Market in order to understand the dynamics of this well-functioning event management model. By publicizing the event, the partners will enhance their reputation and heighten the recognition of the destination.

Discussion and conclusion

This chapter explored whether participants and tourism businesses were psychologically, socially, politically, and economically empowered as a result of the decision-making and implementation processes of the Şile Earth Market. Findings suggest there was variance in the level of empowerment in the four dimensions, and levels of empowerment differed among stakeholder groups. In general, stakeholders were positive about tourism development and happy that visitors to the market were able to experience the town's beauties and to taste traditional foods, that is, they were proud of their heritage (psychological empowerment). This finding corroborates those of Boley and Gaither (2016). The stakeholders also showed awareness of the negative externalities that uncontrolled tourism is creating for their town, solutions to which they expect from the local government. The participants of the Earth Market agreed that the event provided opportunities for them to come together around common goals with other participants. They better connected with other community members thanks to their taking part in the event (social empowerment). This finding agrees with studies that indicate that cultural events have the potential to bring local communities and non-profit organizations together.

Experienced Earth Market participants confirmed they had gained the power to influence the governing committee of the event and that they had become more vocal in communicating their needs (political empowerment).

Findings suggest a high level of interaction and overlap between empowerment dimensions, as discussed in previous research (Moswete and Lacey, 2015). Our findings also demonstrate that, at least in the case of the Earth Market event, one dimension can exist without the presence of the other. For women whose roles are constrained by society's traditional values, participating in the Earth Market event was a mechanism for them to leave behind their homemaking duties in their villages to socialize with their friends and customers in a safe environment without the oversight of family members. This social empowerment is not a direct result of increased economic empowerment; it is merely a result of taking part in the market.

In the Şile Earth Market event, tourism businesses were not included as partners or beneficiaries, nor did they take part in decision-making. The findings reveal that tourism businesses were unaware that this market event could benefit their businesses. Some of them openly criticized the market and discredited the products sold, expressing their disapproval of the event as a whole. This finding supports previous research to the effect that community members often question the benefits and costs of such an event before eventually supporting it. Finally, the findings show that the producers and farmers empowered by the market are more supportive of it as a tourism event than the tourism businesses.

This chapter illustrates that the empowerment of entrepreneurs in the market started a *positive loop* that fosters rural entrepreneurship in Şile. Barriers such as Şile's distance from Istanbul city center had previously hindered rural entrepreneurship from flourishing in Şile. The event provided a safe place where solutions existed. The technical support and resources shared with entrepreneurs have enabled them to take part in the market and overcome some of these barriers. The more ambitious participants have reinvested in their businesses and have created new products and services. This is not to say that barriers have disappeared – a low education level and difficulty in getting bank loans continue to be barriers – but for some, the road to entrepreneurship has become more accessible.

The chapter confirms the importance of building partnerships with stakeholders in tourism planning and implementation (Bramwell and Lane, 2000; Perkins and Khoo-Lattimore, 2019). The partnership between the founding members and the government enabled the realization of the event. The findings confirm the different types of resources shared by non-profits and government at the implementation stage, and the resulting partnership interdependence among them (Austin and Seitanidi, 2012). In the planning of the partnership, non-profit members needed the support of the government for financial, physical, and political resources. As noted by Getz (2008), the non-profits had to give up part of their independence in order to receive support from the government. However, the partners' overlapping interests (Gray and Stites, 2013) and the local government taking more responsibility in the market caused a shift in power over time. Once the market was up and running, the government decided it needed less support from non-profits, the number and frequency of consultations declined. Dredge and Whitford (2011) point out that when there is increased participation from the public sphere, the government uses specialized legislation and politics to curtail the upsurge of specific issues. This result reminds us that it is the government agencies' role to assure the use of democratic processes in sustainable tourism development.

We used the findings from the present research to develop a framework for supporting rural entrepreneurship through a planned event. While adopting a sustainable tourism approach, the framework displays the distinct role of multi-stakeholder partnerships and depicts the community as an essential pillar of tourism events, as suggested by Higgins-Desbiolles (2018). The framework implies the exploration of long-term sustainability impacts for all key stakeholders, including entrepreneurs, tourism businesses, and the founding partners of the event and considers impacts on areas beyond the destination itself.

As for brief suggestions for practitioners, the findings point to the attention necessary by tourism businesses in planning and implementing tourism events. Events founders should find new and better ways to communicate their tourism visions and strategies, the founding principles of the event and the anticipated economic benefits to a larger group of stakeholders in order to gain their valuable support. As for the researchers, one of our presumptions about empowerment and women entrepreneurship in rural areas misled us initially. Women and men alike need their family and community for growth for two reasons. First, entrepreneurs continue to have limited resources for investing in labor (McElwee and Atherton, 2011). Second, the collectivistic culture of the country supports more collectivistic forms of entrepreneurship. Our study revealed that in order to investigate rural entrepreneurship in the context of Turkey, scholars should adopt a broader approach. Rather than examining individual entrepreneurs, co-entrepreneurship and collective forms of entrepreneurship require greater attention.

References

Austin, J. E., & Seitanidi, M. M. (2012). Collaborative value creation: A review of partnering between nonprofits and businesses. *Nonprofit and Voluntary Sector Quarterly, 41*(6), 929–968.

Bazzani, C., Asioli, D., Canavari, M., & Gozzoli, E. (2016). Consumer perceptions and attitudes towards farmers' markets: The case of a Slow Food Earth Market. *Economia Agro-alimentare, 3*, 283–302.

Boley, B. B., & Gaither, C. J. (2016). Exploring empowerment within the Gullah Geechee cultural heritage corridor: Implications for heritage tourism development in the Lowcountry. *Journal of Heritage Tourism, 11*(2), 155–176.

Boley, B. B., & McGehee, N. G. (2014). Measuring empowerment: Developing and validating the resident empowerment through tourism scale (RETS). *Tourism Management, 45*, 85–94.

Boley, B. B., McGehee, N. G., Perdue, R. R., & Long, P. (2014). Empowerment and resident attitudes toward tourism: Strengthening the theoretical foundation through a Weberian lens. *Annals of Tourism Research, 49*, 33–50.

Bramwell, B., & Lane, B. (2000). Collaboration and partnerships in tourism planning. In B. Bramwell, & B. Lane (eds.), *Tourism Collaboration and Partnerships: Politics, Practice and Sustainability* (Vol. 2, pp. 1–19). Celevedon: Channel View Publications.

Butler, G. (2017). Fostering community empowerment and capacity building through tourism: Perspectives from Dullstroom, South Africa. *Journal of Tourism and Cultural Change, 15*(3), 199–212.

Capriello, A., & Rotherham, I. D. (2013). Building a preliminary model of event management for rural communities. In L. Dwyer, & E. Wickens (eds.), *Event Tourism and Cultural Tourism: Issues and Debates* (pp. 8–25). London: Routledge.

Dredge, D., & Whitford, M. (2011). Event tourism governance and the public sphere. *Journal of Sustainable Tourism, 19*(4-5), 479–499.

Getz, D. (2008). Event tourism: Definition, evolution, and research. *Tourism Management, 29*(3), 403–428.

Getz, D., & Page, S. J. (2016a). *Event Studies: Theory, Research and Policy for Planned Events.* London: Routledge.

Getz, D., & Page, S. J. (2016b). Progress and prospects for event tourism research. *Tourism Management, 52*, 593–631.

Gray, B., & Stites, J. (2013). *Sustainability Through Partnerships: Capitilizing on Collaboration.* nbs.net/knowledge: Network for Business Sustainability.

Hatipoglu, B., Aktan, V., Duzel, D., Kocabas, E., & Sen, B. (2016). Developing food tourism through collaborative efforts within the heritage tourism destination of Foça, Izmir. In D. Alvarez, F. Go and A. Yuksel (eds) *Heritage Tourism Destinations: Preservation, Communication and Development.* London: CABI, pp.63–75.

Higgins-Desbiolles, F. (2018). Event tourism and event imposition: A critical case study from Kangaroo Island, South Australia. *Tourism Management, 64*, 73–86.

Lordkipanidze, M., Brezet, H., & Backman, M. (2005). The entrepreneurship factor in sustainable tourism development. *Journal of Cleaner Production, 13*(8), 787–798.

McElwee, G., & Atherton, A. (2011). Rural entrepreneurship. In L. P. Dana (ed.), *World Encyclopedia of Entrepreneurship* (pp. 377–385). Cheltenham: Edward Elgar Publishing Limited.

Moswete, N., & Lacey, G. (2015). "Women cannot lead": Empowering women through cultural tourism in Botswana. *Journal of Sustainable Tourism, 23*(4), 600–617.

Perkins, R., & Khoo-Lattimore, C. (2019). Friend or foe: Challenges to collaboration success at different lifecycle stages for regional small tourism firms in Australia. *Tourism and Hospitality Research, 20*(2), 1–19.

Scheyvens, R. (1999). Ecotourism and the empowerment of local communities. *Tourism Management, 20*(2), 245–249.

Sherwood, P. (2007). *A Triple Bottom Line Evaluation of the Impact of Special Events: The Development of Indicators.* Melbourne, Victoria: Unpublished Doctoral dissertation.

Slow Food. (2020, January 26). *About Us.* Retrieved January 26, 2020, from Slow Food: https://www.slowfood.com/about-us/

United Nations Environment Programme; World Tourism Organisation. (2005). *Making Tourism More Sustainable: A Guide for Policy Makers.* Retrieved from Sustainable Tourism Development: sdt.unwto.org/content/about-us-5

7 Riding wanderlust

The case of motorcycle events

Lucia Cicero

Introduction

According to Sykes and Kelly (2014), the economic impact of motorcycle events in the United States is considerable. Main events and rallies attract hundred thousands of attendees such as the Sturgis Rally, the Daytona races, the Biketoberfest, the Laconia Motorcycle Week, the BBB Rally, the Myrtle Beach Bike (Sykes and Kelly, 2014). In South Africa, thousands of participants attend events such as the Toy Run or Africa Bike Week. In Europe, international motorcycle events, such as the TT race in the Isle of Man (United Kingdom), the Pingüinos (Spain), the Elefantentreffen (Germany) and the European Bike Week (Austria), are powerful attractors for thousands of motorcyclists coming from all around Europe. Apart from big renowned events, moto clubs or associations promote many local events. In general, motorcycle events often last several days and offer a mixture of rallies, commercial stands, food venues, music concerts and entertainment. It follows that motorcyclists move towards motorcycle events to attend a variety of sub-events, where both the travel and the event itself appear as important.

Understanding the tourism motivation concept of wanderlust can explain riders' attitude towards travelling. The conceptualization of wanderlust has facilitated the evolution of motivation theory throughout the last 50 years of research (e.g. Gray, 1970; Crompton, 1979; Leiper, 1989; Figler et al., 1992). Nonetheless, it is rarely considered when outlining travellers' tourism motivation or event attendance motivation.

To investigate the above-mentioned concept in further depth this chapter seeks to answer the questions: how does wanderlust interact with travel attitudes and behaviours of event attendees? The sub-question is: is it meaningful to segment event attendees according to their reported wanderlust? Motorcycle event attendees provide the empirical data set, for its relevance in understanding motivations of riders as people aiming at enjoying the travel, whatsoever the destination. The aim of this chapter is both to explore the concept of wanderlust and motivations of motorcycle event attendees from an event tourism marketing perspective.

Event tourism research appears as multi-faceted and embodied in the framework for studying knowledge on event tourism by Getz and Page (2016). The framework entails the interrelationships among the core phenomenon (i.e. event) and other event-related features, such as its management, the personal antecedents and decision-making, the patterns and processes, as well as the impacts. The review of Getz and Page (2016) leads to the identification of a list of major themes and related future research directions for experience and meaning on event tourism. Nonetheless, it is possible to detect an emerging gap related to motorcycle events.

The main objective of this chapter is to investigate wanderlust as a factor influencing both the travel characteristics and tourism motives of motorcycle event participants. In the case of motorcycle events, attendance could include a short travel of few hours, as well as a combination of travel and event attendance lasting more days. Accordingly, the specific empirical set could include travel combinations both leisure-alike and tourism-alike, depending on the length of the stay.

The theoretical background, which follows, examines the extant literature to help frame former studies related to motorcycle tourism and motorcycle events. Further, this chapter presents the conceptual evolution of wanderlust from the tourism perspective. Among travel motives, wanderlust expresses the will of travel, which intuitively could adhere to the spirit of riders. An initial set of qualitative interview data explored travel attitudes and behaviour related to motorcycle events. Next, in alignment to existing literature and results of preliminary qualitative research, an online self-administered questionnaire addresses motorcycle event attendees. This chapter proceeds with the methods and findings. Subsequently, implications to destination managers and event organizers are proposed. Concluding remarks close this chapter with reflections on limitations and future research indications.

Theoretical background

Motorcycle tourism and tourists

Motorcycle tourism arguably is a form of niche tourism, where the mode of transport identifies key participants. It is a peculiar form of tourism and leisure activity where travelling to the destination tends to be more important than the destination itself (Broughton and Walker, 2009; Walker, 2011; Pinch and Reimer, 2012; Sykes and Kelly, 2014; Cater, 2017). Motorcycle tourism is commonly categorized as a drive tourism typology (Walker, 2011; Cater, 2017), and has unique dimensions such as the attachment to a vehicle and the likelihood of travelling in groups (Walker, 2011; Pinch and Reimer, 2012; Cater, 2017). Motorcycle tourism involves: (i) trips away from home whose main purpose is vacation, leisure, entertainment or recreation; (ii) participation in sport events or the attendance of motorcycling events and rallies; (iii) either active (i.e. driver) or passive (i.e. passenger) tourists; and

(iv) touring motorcycles, i.e. cruisers, touring and sport-tourers (Broughton and Walker, 2009; Sykes and Kelly, 2014).

Further, motorcyclists are recognized for being social and moving in groups, with a strong sense of belonging to the biking community (Broughton, 2007; Austin and Gagné, 2008; Broughton and Walker, 2009; Walker, 2011). Moreover,

> within the biking community there are some that refer to themselves as bikers, others prefer the title of motorcyclists or riders. Which title is used often depends upon the riders themselves; however, the reaction from the public may be different depending on the title.
>
> (Broughton and Walker, 2009: p. 60)

Austin and Gagné (2008: p. 419) cite their collective identity and solidarity as one of the main traits in shaping "a community boundary between riders and non-riders".

Importantly motorcyclists are adventure seekers (Cater, 2017), with a need for constant assessment of risk (Broughton and Walker, 2009; Illum, 2011; Walker, 2011; Terry et al., 2015; Cater, 2017). The study of Price-Davies (2011) aligns the phenomenon of adventure motorcycling with the tourist gaze concept by Urry (1990). Nonetheless, the distinction with adventure tourism is that motorcycle tourism deals with risk, reported about motor-cycle tourism in general requiring awareness and valuation, in spite of active seek of risk (Fuller et al., 2008; Cater, 2017).

The study of Sykes and Kelly (2014) attempts to advance the knowledge of motorcycle tourism from an experiential tourism perspective. Based on the framework for consumer/tourist experience by Ritchie and Hudson (2009), the study of Sykes and Kelly (2014) proposes a new model called Dynamic Interaction Leisure. In the model, interaction and sharing among riders play a pivotal role. Motorcycle tourists interact with the external environment in an intimate way, seek for authentic experiences and expect to regenerate and spiritually renew by motorcycle tourism activities. Besides the model, there is no evidence of studies that examined motivation for motorcycle tourism. However, Sykes and Kelly (2016) focused on destination choice motives by motorcycle tourists.

Motorcyclists appear to have reasonable disposable income, and, as such, represent high revenue potential for destinations (Broughton, 2007; Broughton and Walker, 2009; Walker, 2011; Sykes and Kelly, 2016; Cater, 2017). Nonetheless, a widespread perception of bikers negatively affects the openness by some tourist operators towards this market segment (Broughton and Walker, 2009; Walker, 2011; Sykes and Kelly, 2014; Cater, 2017). On the contrary, biker-focused initiatives have been promoted by specialized op-erators or destinations, taking the form of arranging individual or group motorcycling holidays, club-run tours, guided and organized bike tours, among others (Broughton and Walker, 2009; Walker, 2011; Sykes and Kelly, 2014;

Cater, 2017). Hence from the destination management perspective, motorcycle tourism could represent a key target market for promoting rural destinations especially (Sykes and Kelly, 2016; Cater, 2017). For this purpose, Sykes and Kelly (2016) suggest the creation of apposite Regional Tourism Organizations (RTOs), able to both promote the destination and curtail negative impacts of motorcycle tourism. Their role could be a key one regarding perception of bikers' community and bikers' safety above all (Sykes and Kelly, 2016; Cater, 2017).

Motorcycle clubs play an important role too in developing motorcycle tourism initiatives, including event promotion. Many motorcyclists join motorcycle clubs, which are often far away from the outlaw culture of motorcycle gangs (Broughton and Walker, 2009; Walker, 2011). Nonetheless, such stereotype is often associated with all clubs. In practical terms, clubs may act as promoters of events such as rideouts (i.e. organized runs with pre-determined routes) either on a local basis or with the support of national associations.

Motorcycle events

Motorcycle events may take different forms (Broughton and Walker, 2009; Walker, 2011; Kruger et al., 2014), ranging from small local and short events to large international events. Attending motorcycle events is part of motorcycle tourism. As highlighted by the findings of Cater (2017), over 75% of surveyed motorcycle tourists stated the attendance of motorcycle events: races, rallies, club events, motorcycle shows and legendary events ("must do") such as the Isle of Man TT (for a focus on TT races, see Crowther, 2007 and Terry et al., 2015).

Nonetheless, event tourism research has almost neglected the context of motorcycle events, from a tourism marketing perspective especially. A study as the one by Way et al. (2013), for instance, illustrates demographic characteristics of attendees at Bikes, Blues & BBQ Festival, one of the largest bike rallies in the United States, without focusing on any issue directly related to biker experience. Nale et al. (2003) exclusively focused on the spending behaviour of motorcycle event attendees. Other disciplines contributed to the investigation on motorcycle events, from the standpoint of sociological, anthropological or geographical studies. For instance, Rabinowitz (2007) depicted the celebration of Annual South African Toy Run as a context for expressing biker identity through a multitude of values and styles. Terry et al. (2015) examined spectators' characteristics of TT races at the Isle of Man, finding out that respondents had various attitudes towards risk and revealed a significant female presence. By investigating BMW touring motorcyclists, Austin and Gagné (2008) reported that motorcycle rallies were a fundamental place for biker community formation and maintenance.

From an event tourism perspective, motorcycle events could represent the scenario for investigating attendees through market segmentation according to several characteristics such as demographics, consumer use situation

Table 7.1 Motives for attending motorcycle events

	Cluster 1 "Hardcore bikers"	Cluster 2 "Novice riders"	Cluster 3 "Club cruisers"
Adventure	Extremely important	Important	Very important
Event novelty	Very important	Slightly important	Important
Escape and socialization	Extremely important	Important	Very important
Lifestyle	Extremely important	Slightly important	Very important
Event attractiveness	Extremely important	Important	Very important

Source: Adaptation from Kruger et al. (2014).

or degree of consumer involvement (Assael, 1998). This is the case of Kruger et al. (2014), who examined bikers' motivations for attending a motorcycling event, namely the Africa Bike Week, which attracts bike enthusiasts each year. Their study identified five motivational factors (i.e. adventure, event novelty, escape and socialization, lifestyle and event attractiveness) and detected the presence of three market segments according to the factor combination: Hardcore Bikers, Novice Riders and Club Cruisers.

Table 7.1 illustrates the results of the study by Kruger et al. (2014) in terms of segments and importance of motives for attending a specific motorcycle event (i.e. Africa Bike Week). The analysis of Kruger et al. (2014) represents a great step forward for the comprehension of event tourism motivations in the context of motorcycle tourism, underlining the emergence of differences among motorcycling event attendees, presented as inhomogeneous and requiring ad hoc marketing initiatives. It remains the unique study focused on the motivation for attending a motorcycle event so far. Furthermore, the literature review does not report any study on motives by the entire audience of motorcycle event attendees, regardless of being bikers or not.

The evolving concept of wanderlust in tourism motivation

Gray (1970) suggests wanderlust is juxtaposition to sunlust. He states it represents the will to leave familiar contexts to travel towards the discovery of different cultures and places, with special regards to past cultures and their expressions. In contrast to sunlust, motivation focuses on the available amenities or desired climate conditions of a destination. Later, Gray (1981) reflects on the infrastructures required in order to accommodate and promote wanderlust-driven tourism. In particular, he underlines the greater potential of wanderlust tourism compared to resort tourism in economic terms, despite its cost impacting to a greater extent in infrastructural terms.

Since the works of Gray (1970, 1981), the conceptualization of wanderlust in travel motives has been a complicated issue. Several authors have quoted the contrast between tourism of wanderlust and the one of sunlust (e.g. Crompton, 1979; Leiper, 1989). On one hand, some authors have contemplated wandering as a travel mode rather than wanderlust as a travel motive (Vogt, 1976; Cohen, 1979). On the other hand, Figler et al. (1992) explicitly include wanderlust as a unique motivational factor for pleasure travel. More specifically, they identify five motivational factors for pleasure travel (i.e. anomie/authenticity-seeking, culture/education, escape/regression, wanderlust/exploring the unknown and jetsetting/prestige-seeking). Where wanderlust is associated with exploring the unknown and includes items such as "simply to be on the move", "because of an inexplainable yearning to roam", "to experience danger", "to explore the unknown" (Figler et al., 1992: p. 115). Nevertheless, since the work of Crompton (1979), the relationship between wanderlust and travel motives has been almost neglected by main models on tourism motives (e.g. Iso-Ahola, 1982; Baloglu and Uysal, 1996; Gnoth, 1997; Kozak, 2002; McCabe, 2000; Bond and Falk, 2013), causing a literature gap in the conceptualization of wanderlust from a tourism motivation perspective.

Recently, the use of the term wanderlust has evolved beyond the direction meant by Figler et al. (1992). It has moved to indicate the "impulse to travel" (Shields, 2011: p. 369) rather than a specific type of travel motive or tourism activity. The study of Shields (2011) outlines the role of wanderlust in a scalable way, in order to trace its influence in past, present and expected future travel behaviours. Accordingly, it derives that wanderlust is a travel motivator, which affects travel attitudes and decisions. One of the emerging results of Shields' research is the acknowledgement of wanderlust as a personality trait that characterizes a group of individuals, a willingness to travel that tourism marketers need to address.

Research approach

Data collection

Qualitative interviews with a purposive group of bikers gathered initial insights on the phenomenon. A pilot study developed the main quantitative instrument and included both individual interviews and group discussions. The qualitative approach adopted an informal way and not through a structured form to stimulate respondents in narrating their experiences both as riders and as participants to motorcycle events. Several issues arisen by the interviews later reflect in the questionnaire formulation, in addition to scales provided in the literature. Examples include the belonging to a moto club and the length of staying at motorcycle event according to their distance from home.

To explore wanderlust and event tourism motivations by participants to motorcycle events a wide reaching population was desirable for data collection. For this reason, the design of an online self-administered questionnaire was through Google Forms. This followed the format of Sykes and Kelly (2016) who used Survey Monkey. The first 20 respondents pre-tested the questionnaire, uploaded definitively in March 2018. It required approximately ten minutes for completion. Several Facebook groups of biker communities promoted the survey participation. The result was of 179 collected questionnaires.

Given the exploratory aim of the study and possible cultural differences among consumers with different countries of origin, the questionnaire was written and promoted in a single country (Italy). Questionnaires provided authorization to data treatment according to national law on privacy and complete contact information of academic research unit at starting and ending the questionnaire. The first section of the questionnaire required socio-demographical data (i.e. gender, birth year, education level, job, province of origin).

The second section investigated the experience related to riding a motorcycle such as having the driving license for motorcycle, since when, owning one or more motorcycles and related issues. The decision of going in depth into such questions at the beginning of the questionnaire came to light after the preliminary interviews, when bikers usually attributed great importance to owning a precise type of motorcycle, of a specific brand rather than another and with determinate characteristics in terms of years and power. Acknowledging such importance, in order to break the ice and get closer to bikers' views, the question about the motorcycle type(s) owned was open-ended and categorization was made afterwards.

The third section examined the habits related to motorcycle events, with enquired details on frequency and length of stay at motorcycle events depending on their distance from home (same province, other provinces, other regions, other countries). Information on transportation modes to motorcycle events and their frequency was also required. In the last part of the third section, a question on wanderlust was proposed. The scale on wanderlust is derived by Shields (2011) and Table 7.2 reproduces it.

Table 7.2 Wanderlust scale

Items
I love to pursue new and different vacation experiences
I get really excited waiting for my vacation
I dream about going to exotic travel destinations
I often reflect back on my past vacation experiences
I generally return from a vacation feeling relaxed and happy
I like telling people about the trips I'm planning on taking
I am happiest when I'm on vacation
I like to fantasize about vacation travel
I expect to take a vacation trip at least once a year
Some of my best childhood memories are from vacations I took
5 = high level of agreement, 1 = high level of disagreement.

Source: Shields (2011).

The ten-item scale of Shields (2011) has been adapted to the case of motorcycle event, rewording the sentences when necessary to assign the items to the motorcycle event travel. As in the original scale, respondents were asked to rate the level of agreement with a Likert-scale from one (completely disagree) to five (completely agree). In order to avoid biases, items were in random order to respondents.

The fourth section concluded the questionnaire by investigating the travel motivations for attending a motorcycle event. Given the fact that the literature review on tourism and event motivations is extensive, though with recurrent items, a double selection composed the item group. First, a scale on tourist motivations reflects tourist features in travelling to motorcycle events. For this purpose, an adaptation from Kozak (2002) matched with a second item group, derived by Kruger et al. (2014). Table 7.3 groups both scales.

This second selection included event-specific items related to motorcycle event attendance. Nonetheless, the selection and adaptation of items contemplate general motorcycle event attendance, while in the original form they related to a unique motorcycle event. Respondents were asked to rate the level of importance of proposed items with a Likert-scale from one (not important at all) to five (extremely important). As in the case of wanderlust items, motivational items appeared in random order to respondents.

Data analysis

Data collection occurred through the tools provided by Google Forms and analyzed with IBM SPSS Statistics. Checking for duplicates and internal coherence of each questionnaire determined the exclusion of only one questionnaire, resulting into a final sample of 178 usable responses. The amount is more than satisfactory, considering the results of comparable recent works on motorcycle tourism. Sykes and Kelly (2016) collected 127 survey responses and Cater (2017) 156.

Method

Initial descriptive statistics calculated the sample distribution throughout the inquired data. Afterwards, data treatment traced sample characteristics on event tourism features related to motorcycle events. The next section will describe the results of the empirical analysis, conducted with the use of both principal-component factor analysis to reduce the number of items to a smaller set of factors and cluster analysis to group respondents together according to their attitudes. Finally, several cross-tabs and ANOVA analyses were performed to explore the presence of significant differences among groups of respondents.

Description of the sample

The final sample consists of 178 respondents, of which 77% are males. The average age is 40.62 years old, with a range of respondents from 16 to

Table 7.3 Scales on travel motivation

Tourist motivations	
Factor 1: Culture	To increase knowledge of new places
	To visit historical and cultural sites
	To meet local people
Factor 2: Pleasure-seeking/fantasy	To have fun
	To mix with fellow tourists
	To seek adventure
	To get away from home
Factor 3: Relaxation	To relax
	To be emotionally and physically refreshed
	To enjoy good weather
	To spend time with people cared deeply about
Factor 4: Physical	To engage in sports
	To be active
	To get close to nature

Motives for attending Africa Bike Week	
Factor 1: Adventure	For the adventure of it
	Because of the feeling of freedom associated with the ride
	Because of the thrill of the experience
	To have fun
	To be part of this unique and exciting event
	For a chance to be with people who are enjoying themselves
Factor 2: Event novelty	To support the vendors
	To purchase motorcycle merchandise
	To take part in the Ride-in-Bike show and Mass Ride
	To gain colours/patches
	Because 2013 marks Harley-Davidson's 110th year
	Because of the thrill and rebellious culture associated with motorcycling
	Because I am a Harley-Davidson enthusiast
	Because it is a club event to foster relations with other bikers
	Because of the social status associated with motorcycling
	To share group identity with other bikers
	To meet new people with similar interests
	For nostalgic reasons/memories
	To see a variety of motorcycles up-close
	It is a sociable event
Factor 3: Escape and socialization	To relax
	To get away from my daily routine
	To spend time with family/partners/spouse and friends
Factor 4: Lifestyle	It is part of my lifestyle
Factor 5: Event attractiveness	I attend it annually
	The atmosphere of Africa Bike Week
	Africa Bike Week is one of South Africa's premier motorcycle events
	Because I am a motorcycle enthusiast
	To experience the Africa Bike Week attributes (shows, performances)
	Because the event is well organized

Source: Kozak (2002); Kruger et al. (2014).

71 years old. Table 7.4 illustrates the description of the sample in terms of socio-demographical data. Most respondents hold a high-school diploma (50.3%) or even a Bachelor's/Master's/Postgraduate degree (20.3%) and work as employees/teachers (34.1%) or workers (28.9%). 91.6% of the surveyed people have a driving license for motorcycle, which is compulsory in Italy for driving motorcycles above 125 cc and 11 kW, on average for 16.89 years. 89.9% of respondents own at least one motorcycle and 6.7% more than one. The largest number of frequencies is recorded in the following brands: Harley-Davidson (61 owners), Yamaha (24 owners) and Honda (20 owners). Notably, 31.2% of motorcycle owners bought the first motorcycle before turning 18 years old and overall 65% within the age of 25 years old. One-third of respondents (37.1%) enrolled to a moto club, on average for 8.55 years.

Table 7.4 Sample descriptive statistics

		(%)
Gender	Male	77.0
	Female	23.0
Age	(Mean = 40.62 years old)	
	≤30 years old	18.6
	31–40 years old	36.2
	41–50 years old	24.9
	≥51 years old	20.3
Education		
	Middle school	15.3
	Professional diploma	14.1
	High-school diploma	50.3
	Bachelor's degree	7.9
	Master's degree or postgraduate specialization	12.4
Job		
	Employee/teacher	34.1
	Worker	28.9
	Self-employed	17.3
	Retired	5.2
	Student	4.6
	Executive	3.5
	Other jobs	6.4

Source: Own elaboration.

Findings

Travelling to motorcycle events

As shown in Table 7.5, respondents attend on average 9.20 motorcycle events in one year. Results show that the probability of attending an event decreases depending on its distance, with higher values on closer events (mean score = 3.45) rather than events abroad (mean score = 2.06). Furthermore, the length of the stay varies as well depending on the distance of the event. To attend an event in the same province requires on average half a day or one evening for the majority of respondents (55.2%), while travelling to a motorcycle event in another province tends to require one entire day (53.7% of respondents). Travelling and staying at motorcycle events in other regions or even in other countries means staying away from home more than one day for, respectively, 59.1% and 80% of respondents.

Respondents usually attend motorcycle events with friends or partners. On a scale from one (never) to five (very often), the mean score for friends is 3.99 and for partners 3.05. Infrequent company scenarios are to go alone (2.28) or with family members (2.09). Most respondents (91.9%) travel with motorcycles to go to motorcycle events often or very often. On the same Likert-scale of the previous question, going by riding a motorcycle records a mean score of 4.65, while going by car 1.98 and with other means of transport 1.16. Nonetheless, the former score decreases when respondents are asked to rate the probability of travelling by motorcycle to events in other regions (3.96) or other countries (3.09).

Respondents rated return intention on a scale from one (not likely at all) to five (extremely likely). To return to an event is likely on average, though with a reverse relationship between probability and distance. The closer the event, the more likely for repeat attendance: the mean score for an event in

Table 7.5 Frequency of travelling to motorcycle events and length of stay

		How long does your stay last at a motorcycle event? (More than one answer is possible)		
	Mean score	One night/half day (%)	One day (%)	More than one day (%)
In the same province area	3.45	55.2	46.7	4.2
In other provinces	3.37	31.1	53.7	23.8
In other regions	3.10	16.8	35.6	59.1
In other countries	2.06	8.4	14.7	80.0
Rate scale: 1 = never; 5 = very often				

Source: Own elaboration.

the same province area is 3.55, while it decreases to 3.16 for events in other regions and 2.44 for events in other countries.

Wanderlust and motorcycle events

Table 7.6 illustrates the mean scores recorded for the items of the wanderlust scale applied to the motorcycle events. Item were rated by respondents according to the level of agreement with a Likert-scale from one (completely disagree) to five (completely agree). Items recording a greater level of agreement are: "I generally return from a motorcycle event feeling relaxed and happy" (mean = 3.94), "I expect to attend a motorcycle event at least once a year" (mean = 3.86), "I like telling people about the motorcycle events I'm planning to attendi" (mean = 3.63), and "I dream about going to peculiar motorcycle events" (3.62).

Reliability analysis was conducted on the scale. Cronbach's Alpha is 0.915, where the acceptance level is typically 0.8 (Bryman and Cramer, 2009) or even 0.7 (Brace et al., 2006). Item-total statistics were checked, and no item was deleted. Dimensionality of the scale was examined by checking that the items were loading on a single component. No item reported scores below 0.4 in Component Analysis and the structure of items within the scale was assessed, reporting a KMO of 0.931. Overall, the scale reported a mean score of 35.55, calculated through the sum of the ten item scores as previously done by Shields (2011), showing a minimum overall score of 12 and a maximum overall score of 50. Moreover, to distinguish three groups of respondents depending on their wanderlust (low, moderate, high), as executed in Shield's study (2011), a cluster analysis was performed. The mean values for each level of wanderlust are: low = 23.98, moderate = 34.94, high = 44.79.

Chi-square tests were performed among the three groups of respondents and socio-demographic information. Despite several not significant differences,

Table 7.6 Wanderlust scale for attending motorcycle events

Item	Mean score
I generally return from a motorcycle event feeling relaxed and happy	3.94
I expect to attend a motorcycle event at least once a year	3.86
I like telling people about the motorcycle events I'm planning on attending	3.63
I dream about going to peculiar motorcycle events	3.62
I am happiest when I'm on a motorcycle event	3.56
I get really excited waiting for a motorcycle event	3.55
I love to pursue new and different experiences in motorcycle events	3.51
I often reflect back on my past experiences at motorcycle events	3.48
Some of my best memories are from motorcycle event I attended	3.29
I like to fantasize about travelling to motorcycle events	3.12
Cronbach's Alpha = 0.917	

Source: Own elaboration.

Table 7.7 Respondents according to wanderlust segments and results of significant tests

	Low n = 47	Moderate n = 64	High n = 63	
Mean wanderlust score	23.98	34.94	44.79	F = 575.233 Sig. < 0.001
≤30 years old	15.2%	48.5%	36.4%	100%
31–40 years old	33.3%	27.0%	39.7%	100%
41–50 years old	21.4%	33.3%	45.2%	100%
≥51 years old	34.3%	48.6%	17.1%	100%
Probability of attending motorcycle events in other countries (mean score) (1=not likely at all; 5=extremely likely)	1.66	2.09	2.39	F=6.000 Sig. = 0.003
Frequency of travelling to motorcycle events by car (1=never; 5=very often)	1.47	1.94	2.41	F = 10.780 Sig. < 0.001
Travel to motorcycle events: only by motorcycle	Yes = 71%	Yes = 42%	Yes = 29%	χ^2 = 19.486 Sig. < 0.001
Frequency of travelling to motorcycle events with own moto club mates	1.89	2.30	2.98	F = 6.249 Sig. = 0.002
Frequency of travelling to motorcycle events with members of other moto clubs (1=never; 5=very often)	1.76	2.42	2.79	F = 9.838 Sig. < 0.001
Probability of revisit motorcycle events in same province (mean score)	3.07	3.53	3.90	F = 12.564 Sig. < 0.001
Probability of revisit motorcycle events in other provinces (mean score)	2.89	3.34	3.89	F = 19.758 Sig. < 0.001
Probability of revisit motorcycle events in other regions (mean score)	2.47	3.20	3.60	F = 18.408 Sig. < 0.001
Probability of revisit motorcycle events in other countries (mean score) (1=not likely at all; 5=extremely likely)	1.71	2.41	3.00	F = 13.553 Sig. < 0.001

Source: Own elaboration.

such as in the case of gender, age was reported as significantly different among the three groups (χ^2 = 12.813; sig. = 0.046). Table 7.7 includes the composition of the groups by age of respondents. The lowest level of wanderlust is in the eldest age class (≥51 years old). Comparatively, the youngest respondents are those with less numerosity within the low wanderlust group. Looking at the probability of attending motorcycle events, significant differences are recorded in the events that take place in other countries (χ^2 = 6.000; sig. = 0.003).

The frequency of travelling to motorcycle events by car is significantly different among the three groups (χ^2 = 10.780; sig. < 0.001). Despite travelling to events by motorcycle is not significantly different among clusters, a significant difference is detected in the choice of travelling only by motorcycle to motorcycle events (χ^2 = 19.486; sig. < 0.001). This choice is more frequent in the low wanderlust group compared to other groups. Travelling with members of one's own or of other moto clubs is also significantly different, recording greater frequency with those grouped in the high wanderlust level. Finally, a great difference lies in the probability of returning to a motorcycle event. Here, independently from the distance of the event, the group with the highest wanderlust level shows the highest probability values, which significantly differ from the ones of the other two groups (all items with sig. < 0.001).

Travel motivations to motorcycle events

Although the primary focus of this study is to outline differences among segments of event attendees based upon a measure of their wanderlust, the motivation to attend motorcycle events is also examined through a pool of motivational items taken and adapted from the studies of Kozak (2002) and Kruger et al. (2014). Table 7.8 illustrates the first five motives according to the overall sample and to each segment.

Table 7.8 Main motives for attending motorcycle events: mean scores (ranks)

	Overall sample	Low n = 47	Moderate n = 64	High n = 63	
To have fun	4.32	3.81	4.30	4.73	F = 22.015
	(1)	(1)	(1)	(1)	Sig. < 0.001
The feeling of freedom	4.19	3.72	4.14	4.59	F = 12.208
associated with the ride	(2)	(2–3)	(2)	(2)	Sig. < 0.001
To know new places	4.14	3.72	4.08	4.51	F = 10.662
	(3)	(2–3)	(3)	(4)	Sig. < 0.001
The atmosphere of the event	4.05	3.47	4.00	4.54	F = 23.520
	(4)	(6)	(4–5)	(3)	Sig. < 0.001
To spend time with people cared	3.94	–	4.00	4.46	F = 25.992
deeply about	(5–6)		(4–5)	(6)	Sig. < 0.001
To be with people who are	3.94	–	–	–	F = 19.201
enjoying themselves	(5–6)				Sig. < 0.001
To enjoy the good weather	–	3.62	–	–	F = 4.034
		(4)			Sig. = 0.019
Being a motorcycle enthusiast	–	3.53	–	–	F = 5.964
		(5)			Sig. = 0.003
The well-organized event	–	–	3.92	–	F = 6.424
			(6)		Sig. = 0.002
To relax	–	–	–	4.49	F = 28.491
(1=not likely at all; 5=extremely likely)				(5)	Sig. < 0.001

Source: Own elaboration.

In the first instance, main motives appear as all significantly different among clusters. Nonetheless, the group with a high wanderlust level always has a comparatively higher value with respect to the other two groups. To examine the relative importance of each motive to each cluster, the rank of motives calculated respondents' attitude towards motorcycle events. Having fun and the feeling of freedom associated with the ride are main motives for the overall sample (respectively, mean = 4.32 and 4.19) and for each cluster (Low wanderlust: having fun = 3.81 and freedom = 3.72; Moderate wanderlust: having fun = 4.30 and freedom = 4.14; High wanderlust: having fun = 4.73 and freedom = 4.59).

To know new places follows, with its relative decreasing importance according to increasing levels of wanderlust scores (Low wanderlust: mean = 3.72, rank = 2–3; Moderate wanderlust: mean = 4.08, rank = 3; High wanderlust: mean = 4.51, rank = 4). Conversely, event atmosphere has relative increasing importance according to increasing levels of wanderlust scores (Low wanderlust: mean = 3.47, rank=6; Moderate wanderlust: mean = 4.00, rank = 4–5; High wanderlust: mean = 4.54, rank = 3).

Enjoying the good weather (Low wanderlust: mean = 3.62, rank = 4) and to be motorcycle enthusiasts (Low wanderlust: mean = 3.53, rank = 5) are likely motives for the group with a low wanderlust level. To enjoy the company of people one cares about is a likely motive for both the moderate and high wanderlust group (Moderate wanderlust: mean = 4.00, rank = 4–5; High wanderlust: mean = 4.46, rank = 6), which also aims at relaxing at a motorcycle event (High wanderlust: mean = 4.49, rank = 5).

Discussion

These findings are analysed against the results emerging from past literature in order to trace thoughts on wanderlust and on motorcycle events. As suggested before, a scarce evolution has characterized the analysis of the wanderlust concept. This study had the goal to advance the knowledge on the wanderlust effect within the context of motorcycle events. Motorcyclists reported high levels of wanderlust, but with significant differences among them. As in the case of Shields (2011), wanderlust has been the criterion to segment motorcycle event attendees into three clusters. Such groups showed different travel preferences, as in the case of those with high wanderlust who travel even by car to attend motorcycle events, and diverse travel attitudes, as recorded in the degree of importance of motives for attending a motorcycle event. Getz and Page (2016) highlighted the potential segmentation of event tourism market according to intrinsic and extrinsic motivators; the present study showed how groups move by different motives, which do not have the same importance for everyone. Despite having fun and feeling the freedom of the ride were the most important items for the overall sample and for each cluster, other motives, such as the discovery of new places or the atmosphere of the event, recorded different results among clusters. For instance, relaxing was one of the main motives only for those with high levels of wanderlust.

Inspecting the overall sample, the percentage of female respondents (23%) signed an evolution of motorcycle event attendees, often dominated by masculinity (Pinch and Reamer, 2012; Terry et al., 2015), although even Cater (2017) reported an increase in female presence. Differently from the study of Cater (2017), the present study recorded a lower mean age, namely around 40 years of age for this study and over 45 for Cater's one.

What assumes great importance from a tourism event perspective is to confirm the importance of meeting and sharing within the biker community, recalling the idea outlined by Gezt and Page (2016) that *communitas* and sharing among participants could be a motive for event travel. The findings show that motorcycle events contribute to the tourism development from an economic point of view as in past studies. As a powerful attraction for an entire community of motorcycle enthusiasts, motorcycle events also represent the meeting point for them, making bikers move towards several events a year to meet other bikers and to share the experience of travelling and enjoying themselves jointly. From a destination marketing perspective, the presence of bikers in a place could be the occasion to host them in local accommodation venues and to promote tourist attractions, as well as local businesses.

The findings also highlight how the event organization should satisfy different bikers' preferences and motives. The atmosphere of the event is one of the main motives for attending a motorcycling event, implying that event organizers should aim at creating the proper environment for such atmosphere, year after year, considering bikers as a naturally loyal market. In order to do so, the organization should guarantee the right balance between fun and safety. For Dredge and Whitfors (2011), major tourism events require active support from actors of public spheres. For this reason, event organizers should aim at establishing a connection with the local public administration to facilitate the promotion and the development of motorcycle events.

Tourism or event marketing researchers could draw significant conclusions in order to reduce the gap on both wanderlust and motorcycle event attendees. The study indeed showed a concrete application of wanderlust analysis by adopting methodological innovation, such as studying the attendance to motorcycle events in itself, despite focusing on a specific event. The reason for this choice derives from extant literature on motorcycle tourism, where the importance of travelling recalled the wanderlust concept, and where the habit of motorcycle event attendance was an emerging aspect of bikers' ordinary life. Hence, the final contribute of the study is to provide views on under-researched areas, tracing the paths of wanderlust effect on motorcycle travel attitudes and behaviours.

Conclusions

This chapter aims at uncovering the impact of wanderlust by motorcycle event attendees, revealing how the high will of attending a motorcycle event, whatsoever the event, could be a criterion for segmenting the biker community. Findings showed a variety of travel preferences depending on wanderlust level and provided a link between theoretical background and motorcyclists' practices.

The design of the empirical research presents some limitations. The sample is limited to a national case and collected questionnaires are not numerous. Nonetheless, comparable recent studies have reported similar or even minor numbers in survey to motorcyclists (Sykes and Kelly, 2016; Cater, 2017). Arguably, the practical investigation on similar issues related to motorcycle tourism through quantitative tools, such as the online self-administered questionnaire, tends to show relatively low response rates.

Nonetheless, further research could aim at extending the scope of surveys towards several national contexts, for instance by starting with countries sharing borders with Italy, and thus more likely to attend similar events. Another way to collect questionnaires could be to conduct the survey at a specific event. Although this could lead to increased respondents gathered questionnaires, it brings in a self-selection bias. As mentioned earlier motorcycle events are various and specific kinds of events that could attract mainly a specific audience within the broader segment of bikers (e.g. mainly Harley-Davidson owners). Future research could apply a combination of mixed-methods, in order to integrate both quantitative and qualitative data collection within the same sample. Furthermore, the writer is currently engaged in a focused analysis of tourism motivations by motorcycle event attendees to advance findings from the study. Finally, further research should deepen the knowledge on travel motivations by motorcycle event attendees in order to embrace an overall shade on motorcycle event movements.

Acknowledgements

I wish to acknowledge the help provided by participants to preliminary studies, who assisted me in discovering a world made of motorcycle brands, technicalities, events, stereotypes, recurrent attitudes and behaviours. A world made of proud bikers. Thank you!

References

Assael, H. (1998). *Consumer Behavior and Marketing Action* (6th Edn.). Cincinnati, OH: South-Western College Publishing.

Austin, D. M., & Gagné, P. (2008). Community in a mobile subculture: The world of the touring motorcyclist. In Denzin, N. K. (Ed.), *Studies in Symbolic Interaction*, Volume 30, 411–437. Urbana, IL: Emerald Group Publishing Limited.

Baloglu, S., & Uysal, M. (1996). Market segments of push and pull motivations: A canonical correlation approach. *International Journal of Contemporary Hospitality Management, 8*(3), 32–38.

Bond, N., & Falk, J. (2013). Tourism and identity-related motivations: Why am I here (and not there)? *International Journal of Tourism Research, 15*(5), 430–442.

Brace, N., Kemp, R., & Snelgar, R. (2006). *SPSS for Psychologists* (3rd Edn.). Hampshire: Palgrave Macmillan.

Broughton, P. S. (2007). *Risk and Enjoyment in Powered Two Wheeler Use* (Doctoral dissertation, Edinburgh Napier University). Boca Raton, FL

Broughton, P., & Walker, L. (2009). *Motorcycling and Leisure: Understanding the Recreational PTW Rider.* CRC Press.

Bryman, A., & Cramer, D. (2009). Quantitative Data Analysis with SPSS 14, 15 and 16: A Guide for Social Scientists. Hove, ESX: Routledge.

Cater, C. I. (2017). Tourism on two wheels: Patterns of motorcycle leisure in Wales. *Tourism Management, 61,* 180–189.

Cohen, E. (1979). A phenomenology of tourist experiences. *Sociology, 13*(2), 179–201.

Crompton, J. L. (1979). Motivations for pleasure vacation. *Annals of Tourism Research, 6*(4), 408–424.

Crowther, G. (2007). Embodied experiences of motorcycling at the Isle of Man TT Races. *International Journal of Motorcycle Studies, 3*(3). Available at: http://ijms.nova.edu/November2007TT/IJMS_Artcl.Crowther.html

Dredge, D., & Whitford, M. (2011). Event tourism governance and the public sphere. *Journal of Sustainable Tourism, 19*(4–5), 479–499.

Figler, M. H., Weinstein, A. R., Sollers, J. J., & Devan, B. D. (1992). Pleasure travel (tourist) motivation: A factor analytic approach. *Bulletin of the Psychonomic Society, 30*(2), 113–116.

Fuller, R., Hannigan, B., Bates, H., Gormley, M., Stradling, S., Broughton, P., et al. (2008). Road Safety Research Report No 94. Understanding inappropriate high speed: A qualitative analysis (Vol. 94). London: Department for Transport.

Getz, D., & Page, S. J. (2016). Progress and prospects for event tourism research. *Tourism Management, 52,* 593–631.

Gnoth, J. (1997). Tourism motivation and expectation formation. *Annals of Tourism Research, 24*(2), 283–304.

Gray, H. P. (1970). *International Travel: International Trade.* Lexington, Heath Lexington Books.

Gray, H. P. (1981). Wanderlust tourism. Problems of infrastructure. Annals of Tourism Research, 8(2), 285–290.

Illum, S. F. (2011). Motorcycling and leisure: Understanding the recreational PTW rider. Paul Broughton, linda walker. Book review. Tourism Management, 32, 458.

Iso-Ahola, S. E. (1982). Toward a social psychological theory of tourism motivation: A rejoinder. *Annals of Tourism Research, 9*(2), 256–262.

Kozak, M. (2002). Comparative analysis of tourist motivations by nationality and destinations. *Tourism Management, 23*(3), 221–232.

Kruger, M., Viljoen, A., & Saayman, M. (2014). What drives bikers to attend a motorcycling event? *African Journal of Hospitality, Tourism and Leisure, 3*(1), 1–21.

Leiper, N. (1989). Tourism and gambling. *GeoJournal, 19*(3), 269–275.

McCabe, A. S. (2000). Tourism motivation process. *Annals of Tourism Research, 27*(4), 1049–1052.

Nale, R. D., Rauch, D. A., & Barr, P. B. (2003) *A Note on the Economic Implications of a Focused Tourism Event: Bikers in Myrtle Beach.* Available at: http://citeseerx.ist.psu.edu/viewdoc/download?doi=10.1.1.522.7976&rep=rep1&type=pdf

Pinch, P., & Reimer, S. (2012). Moto-mobilities: Geographies of the Motorcycle and Motorcyclists. *Mobilities, 7*(3), 439–457.

Price-Davies, E. (2011). Adventure motorcycling: The tourist gaze. *International Journal of Motorcycle Studies*, 7(1). Available at: http://ijms.nova.edu/Spring2011/IJMS_Artcl.PriceDavies.html.

Rabinowitz, I. (2007). A Generous imaginary: Contingencies of value in the South African charity run. *International Journal of Motorcycle Studies*. Available at: http://ijms. nova. edu/March2007/IJMS_Artcl. Rabinowitz. html

Ritchie, J. R., & Hudson, S. (2009). Understanding and meeting the challenges of consumer/tourist experience research. *International Journal of Tourism Research*, 11(2), 111–126.

Shields, P. O. (2011). A case for wanderlust: Travel behaviors of college students. *Journal of Travel & Tourism Marketing*, 28(4), 369–387.

Sykes, D. M., & Kelly, K. G. (2014). Motorcycle tourism demand generators and dynamic interaction leisure. *International Journal of Culture, Tourism and Hospitality Research*, 8(1), 92–102.

Sykes, D., & Kelly, K. G. (2016). Motorcycle drive tourism leading to rural tourism opportunities. *Tourism Economics*, 22(3), 543–557.

Terry, A., Maddrell, A., Gale, T., & Arlidge, S. (2015). Spectators' negotiations of risk, masculinity and performative mobilities at the TT races. *Mobilities*, 10(4), 628–648.

Urry, J. (1990). Tourist Gaze: Leisure and Travel in Contemporary Societies (Theory, Culture & Society). London: Sage publications.

Vogt, J. W. (1976). Wandering: Youth and travel behavior. *Annals of Tourism Research*, 4(1), 25–41.

Walker, L. (2011). Tourism and leisure motorcycle riding. In: Prideaux, B., & Carson, D. (eds.). *Drive Tourism: Trends and Emerging Markets*. Abingdon, OX: Routledge, 146–158.

Way, K. A., & Robertson, L. J. (2013). Shopping and tourism patterns of attendees of the bikes, blues & BBQ Festival. *Journal of Hospitality Marketing & Management*, 22(1), 116–133.

Part III

Regeneration, displacement and planning frameworks

8 Quake aftermath and conference industry transformation, Christchurch

Abrar Faisal, Julia N. Albrecht and Willem J. L. Coetzee

Introduction

The socio-political consequences of natural disaster-induced crises (Faulkner, 2001; Huan & Shelby, 2006; Henderson, 2007; Pforr, 2009; Ritchie & Campiranon, 2015; Hall, Prayag, & Amore, 2018; Faisal, 2019) inevitably challenge the complex environment of contemporary event industry in global cities. The conference industry is especially vulnerable due to the multidimensional consumption process and its inseparability from the built environment.

The Canterbury earthquakes on February 22, 2011, irrevocably transformed New Zealand's (NZ) second largest city Christchurch's urban landscape and economic geography. The negative economic impacts in the quake-torn environment caused a drastic fall in Christchurch's share of NZ's convention industry. Irreversible damage to the convention facilities and more than half of the building stock, critical infrastructure collapses and long-term closure of the city centre triggered the immediate decline in national conference market share from 25% to only 1% (Christchurch and Canterbury Tourism (CCT), 2011). NZ's most expensive and biggest urban renewal project, central government interventions and entrepreneurial strategies in the market-led economy have remarkably altered Christchurch's socio-political and built environment after the 2010/2011 destructive earthquakes. Tourism-driven economic recovery policy for the city centre and urban renewal anchor projects, comprising the under construction mega convention centre and other venues across the city, reflect the strategic relevance of the conference industry to revive Christchurch.

As Christchurch continues to recover, the conference industry rebounded to 9% of the national market in 2016 (CCT, 2016). This chapter critically examines the Christchurch city centre convention sector during the earthquake aftermath, and it portrays the recovery trajectory of the post-quake conference industry. A synthesis of the literature, as well as industry insights, offers a comprehensive understanding of the critical challenges and opportunities of conference portfolio management in a post-disaster environment. The first section of the chapter surveys the literature on urban regeneration and conference industry. After an explanation of the context and methodological approach, the following sections focus on the rise of an altered city and related governments responses and legislation. Finally, the conference industry is examined as a potential regenerator in a post-disaster city.

Rise of the conference industry in global cities

In search of alternative means of economic development, cities embrace tourism as an entrepreneurial strategy, resulting in major investments in the built environment to recast the city centre as a premier location for businesses, residents and tourists (Spirou, 2011). Reviewing the business event literature, Getz (2008) defines business event tourism as a mixed-motive phenomenon (a mix of business and pleasure travel) and notes the trend of government's increasing awareness of economic benefits of event tourism. In organisational ecology perspective, 'event populations' can be defined as the normative order (by policymakers and planners in collaboration with other stakeholders) of common activities and similar patterns of resource utilisation (Baum, 1996; Hannan, Polos & Carroll, 2007; Getz, 2017).

Conferences are different facets of the same dynamic; similar objectives, activities and resource dependencies of conference organisers and delegates fit into a 'collective identity' (Hannan et al., 2007). Stakeholder relationships, resource availability, product demand and marketing ownership are the key factors to govern destination competency and population dynamics in the conference industry (Rogers, 2013). Getz (2017) suggests the study of whole event population to understand the phenomenon of ecosystem evolution in a given area. Conference population is one of the discrete sectors of event tourism, which is closely aligned to the 'business travel' (Getz, 2008). Being at the high-yield end of the tourism spectrum, conference 'delegate spends' is considered as complementary to other tourism businesses in the destination (Rogers, 2013).

Conference portfolios serve as image-maker for a destination, a catalyst for urban renewal (Getz, 2008) and 'urban boosterism' (Hiller, 2000). The literature defines an event portfolio as a 'leverageable resource', which is the strategic patterning of events in pursuit of multiple goals (Chalip, 2006; Patterson & Getz, 2013), "the dominant one being economic development" (Getz, 2008, p. 422). In collaboration with the private sector and regional tourism organisations, the local government manages the event portfolio to develop and promote tourism. Threats and opportunities are perceived differently through the life cycle stages of event portfolio, and management strategies change as it evolves (Getz, 2017).

In addition to the slow-burning stresses like insufficient infrastructure and socio-political unrest, urban areas are vulnerable to the external shocks of natural disasters. The frequency and intensity of natural disasters have been increasing over the years (Swiss Re, 2017). Disasters have immediate, catastrophic impacts on conference demands, leading to cancellation or postponement of conferences scheduled in the affected region. The prolonged disaster-induced crisis may cause the extinction of the partial or entire conference population. Subsequently, resource constraints and risk perception of the impacted region can also stimulate conference demand in competing destinations (Rogers, 2013; Getz, 2017).

A significant component of any event is the experiences created by the event manager. Rittichainuwat and Mair (2012) argue that business events (conferences, trade shows) possess both extrinsic and intrinsic motivation (leisure desires). In the urban context, Getz (2017) notes the imperatives of a supportive environment (policy and strategy, legislation, resources) for sustainable event portfolios in pursuit of socio-economic development of the destination. Faisal, Albrecht and Coetzee (2017) argue that post-disaster chaos and the rebuild phase offer opportunities of (re)gaining sustainable business proposition for long-term development, plans and actions, which could not have existed before. Anderssson and Getz (2009) apply institutional theory and posit institutional embeddedness, committed stakeholders and resource dependency as the critical issues of institutionalisation of an event population in any given environment. The policy domain of an ecosystem permanently supports the high-value population in event tourism. Some of the institutionalised events or event portfolio (either because of strategy or by an evolutionary process) may survive indefinitely within a selective environment, "as an asset to cultivate symbiotically with others for long-term, sustainable value" (Getz, 2017, p. 582).

Tourism-based urban regeneration strategies often tend to transform the leisure and tourism spaces (Spirou, 2011), often tied with the event tourism elements, such as convention centres, cultural venues, sports facilities and exhibition centres (Amore & Hall, 2016). As a spatial-temporal phenomenon (Getz, 2008), the supply side of the conference industry has historically been driven by the needs of trade and commerce (Rogers, 2013), and evolved as the vital economic benefactor and regeneration catalyst for urban destinations (Karski, 1990). Contemporary literature on urban tourism and urban regeneration addresses the socio-economic dimensions of visitor economy (Maitland, 2009; Ashworth & Page, 2011; Spirou, 2011) and extend interdisciplinary scholarly discussions on (re)creating events, tourism and leisure spaces to decisively alter the trajectory of urban (re)development. However, Amore and Hall (2016) argue that the phenomenon of urban regeneration in the context of post-disaster recovery is overlooked within the discourses of post-disaster recovery. They examine the traits of both the mainstream and post-disaster regeneration and state that both forms of urban regenerations are "conceived as an opportunity to radically change the urban landscape and tackle perceived economic decline" (p. 182). Stevenson, Seville, Kachali, Vargo and Whitman (2011) emphasise the evolving networks of the environmental forces and explain that the linkages among organisations and their social, economic and physical environment are decisive in organisational recovery and resilience. Their study in the context of post-disaster Christchurch suggests a collaborative approach to reduce individual vulnerability, by partnerships and networks with other organisations in the given environment. This study operationalises the socio-political, legal and macroeconomic forces as non-modifiable environment conditions, and adopts the theoretical lens of organisational ecology to present, interpret and analyse contextual responses and collective efforts of key stakeholders to manage conference portfolio during the aftermath of a major natural disaster.

Context and methodological approach

Christchurch is the South Island's largest and NZ's second largest city. As a major strategic transport hub and financial centre with one of the biggest seaports and the second largest airport, Christchurch is the tourism gateway to South Island. The city was home to 377,000 people in June 2010 – about 8.5% of the country's total population. Before 2010, it had a crucial economic role for the country. The city of Christchurch generated over 70% of the Canterbury region's economic output accounting 10% of national Gross Domestic Product (GDP) (Canterbury Earthquake Recovery Authority (CERA), 2012a; Christchurch City Council, 2011).

Historically, the area inside Bealey Avenue, Fitzgerald Avenue, Moorhouse Avenue and Deans Avenue has been the business and socio-cultural heart of the city (Christchurch City Council, 2011), comprising a surface area of 6.33 square km for the Christchurch Business District (CBD). This land is known to be highly unstable geologically and relatively close to the major seismic activities along the Alpine Fault (Orchiston, 2012). A large regional network of services and infrastructure is supported by the CBD. It was the home of 8,000 people and 6,000 businesses employing 51,000 professionals before the earthquake sequence (Greater Christchurch Group, 2017).

Tourism contributed 8% of the regional GDP and 11% of the employment. Canterbury's share of international arrival was 22% in 2010, supported by the 5.9% annual growth in Christchurch, more than the national growth of 3.8%. At the time the central city attracted 1.8 million visitors annually, who spent more than NZ\$ 2.7 billion in the Canterbury region. Australian visitor arrivals, the key market, increased by 11.1% in the region and 10% in the city, whereas the number of Chinese visitors decreased by 2.2% in the region and increased by 21.2% in the city. Both the Canterbury region and Christchurch experienced a sharp decline in the Japanese market; it used to be 5.5% in the region and 19.4% in the city. The domestic market of the city was in stagnation, with the slow growth of domestic visitors by 0.9% (all data CCT, 2011).

A series of more than 11,000 earthquakes registered between 2010 and 2011 changed Christchurch's landscape forever (CERA, 2014). The earthquakes and aftershocks sequence directly affected 460,000 people in the Canterbury region, resulting in losses of 185 lives, and serious injury, stress and mental trauma for large parts of the population in greater Christchurch (Greater Christchurch Group, 2017). The earthquake of February 22, 2011, was the most destructive natural disaster in NZ since the Napier earthquake of 1931. The earthquakes and aftershocks continued for more than a year, differentiating Christchurch from other global cities that were struck by a single event. These unprecedented earthquakes have had far-reaching socio-economic consequences in NZ's recorded history causing the most expensive and biggest urban renewal initiatives (Greater Christchurch Group, 2017). Following the first earthquake on the September 4, 2010, the Canterbury region experienced over 13,000 aftershocks by mid-2014. The damages to the built environment were estimated at NZ$ 40 billion (NZ Trade and Enterprises, 2016), making it the costliest natural disaster in NZ to date, representing around 20% of the NZ's annual GDP (CERA, 2016b). It also caused a downgrade in long-term sovereign rating (by Standard and Poor's) of NZ (Brookie, 2012).

The quakes continue to significantly disrupt the Christchurch tourism sector suffering from capacity loss. The related identity crisis also needs to be addressed in the restoration (Stevenson et al., 2011). The city had 3.2 million international visitor nights a year which plunged by one million following the earthquake disaster (Bradley, 2013, August 28). Before the earthquake, the estimated revenues of the Christchurch conference and convention industry were NZ$ 100 million a year, having 55.66% of the international delegate days in NZ (CCT, 2010, 2011).

This study qualitatively analyses the literature on event tourism, focusing on the conference industry, in the context of the post-disaster urban regeneration process. The synthesis of the event tourism and disaster impacts discourse addresses the complex issues in the post-quake Christchurch conference industry. The discussion is based on more than 70 published documents like scientific papers, industry reports, news articles, policy documents, official statistics and reports of the NZ government on impacts of the Christchurch earthquakes, specifically the complexities in the recovery of the city's built environment and visitor economy. Drawing on the insights of the post-quake governance arrangements, organisational responses, recovery legislation and regeneration strategy, the trajectory of the conference industry portfolio is critically examined in the tourism ecosystem perspective. The emerging themes from the systematic and comprehensive literature review identify the critical challenges of recovery from large-scale disaster. The analysis phase adopts the theoretical lenses of 'organisational ecology' (Hannan et al., 2007) to generate an in-depth understanding of the findings and construct knowledge of business continuity strategies in environmental uncertainties.

The quake effects in Christchurch CBD

The Christchurch earthquakes on February 22, 2011

> have caused a ground surface rupture, ground shaking, liquefaction, lateral spread, rock fall, topographic amplification, landslides, regional

uplift and subsidence, ground compaction and ground surface renting. The Central City experienced severe ground shaking, and liquefaction and lateral spreading, and subsidence occurred in the north-east part of the city.

(CCC, 2011, p. 22)

The damage and demolition sustained to 1,628,429 square metres of road (52% of Christchurch's sealed roads), 124 km of water mains, 529 kilometres of sewer pipes, 168,000 dwellings (representing 90% of the Christchurch's housing stocks) and 1,100 buildings in the city centre (CERA, 2013, 2015; Greater Christchurch Group, 2017; NZ Trade and Enterprises, 2016).

The Christchurch CBD has been the historic centre of the city; more than half of the 250 listed heritage buildings were located in the central city. A total of 173 sites in the city centre refer to the buildings and places of historical and cultural significance (Amore, 2016), most of which were built with unreinforced masonry techniques not suited for earthquake-prone areas (Ingham & Griffith, 2010). About 113 of the severely damaged central city heritage buildings were demolished by the end of 2011, presenting a significant challenge in retaining the city's links to the past. City centre businesses were dealing with the 'new normal', resulting from the widespread damage of the built environment, long-term cordons of the city centre, changes in regulatory policies and changes in customer demography. The CBD population fell by 35.9% to 4,902, several businesses plummeted to 3,758 and the number of employees dropped to 27,560 (Greater Christchurch Group, 2017).

Importantly, Christchurch CBD was the home of NZ's only purpose-built convention centre and South Island's largest in-house hotel conference centre at the Hotel Grand Chancellor. Both facilities were severely damaged beyond repair. The Christchurch Conference Bureau (CCB) reported an estimated loss of NZ$ 250 million of conference and convention businesses, given the limited capacity of hosting the large conferences (CCT, 2013; Table 8.1).

Earthquake damages reduced the number of hotel rooms by 78%, and it was a 43% loss in hosting capacity. The national share of Christchurch's Tourism Industry dropped from 9.2% pre-earthquake to 6.9% in 2012 (CCT, 2013). Tim Hunter, the chief executive of CCT, notes that "there is no event

Table 8.1 Christchurch venues and conference facilities in hotels

Pre-February 22, 2011		Post-February 22, 2011	
Functional	*Hosting capacity*	*Functional*	*Hosting capacity*
27 Hotels	5,665	11 Hotels	1,935
15 Venues	16,942	9 Venues	11,010

Source: Christchurch and Canterbury Tourism (2011).

since World War II that has disrupted our tourism industry by so much and for so long" (Bradley, 2013, August 28). Most of the tourism businesses, including the accommodation sector and SMEs, located near to the CBD were badly affected than those located in the periphery region (Orchiston, 2012); commercial accommodation units ave lable in the central city fell from 5,279 in March 2010 to 621 in March 2012 (Tourism Ticker, 2017, June 12).

Many businesses relocated and dispersed in the city edges. Moreover, the CBD was the transport hub of the city; the closure caused a significant disruption in the transport network (Greater Christchurch Group, 2017). Changes in consumer demography altered the distribution of consumers and the workforce, resulting in a shift from East to West. Relocations of central city businesses to commercial land in the west of the city were the most significant contributor to reversing the historic trends of economic activities concentration in the central city (CERA, 2013b).

Recovery and rebuild: reframing the urban spaces

The Christchurch Central Recovery Plan identifies the Christchurch CBD as a key strategic centre of national economy and tourism sector. The recovery blueprint focuses on the development and delivery of 17 anchor projects in the proposed precincts within a compressed CBD, mostly the tourism and recreational attractions and facilities, civic buildings, and redesigned city blocks and transport networks to stimulate the redevelopments (CERA, 2012a). Arguably, the multi-billion-dollar anchor projects aim to redefine the city, recreate city spaces and provide certainty for the businesses, residents and visitors.

The total rebuild cost including betterment is estimated at NZ$45 billion; the private sector and Christchurch City Council (CCC) are contributing over half of the rebuild spend through insurance and investment. Central government's investment in response to the Canterbury earthquake is NZ$ 17.5 billion to date (CERA, 2016b; NZ Trade and Enterprises, 2016).

Recovery governance and legislation

New governance structures evolved in response to the Canterbury earthquake sequence. An emergency response took place immediately after the September 2011 earthquake, under the Civil Defence Emergency Management framework (Civil Defence Emergency Management Act 2002). But the institutional design and statutory framework were inadequate for long-term recovery (Brookie, 2012). It became the impetus to review the framework and improve the legislative framework to deal with the Canterbury earthquake sequence. Following the September 2010 earthquake, NZ's Prime Minister appointed a dedicated Minister for Earthquake Recovery. The powers available under the Civil Defence Emergency Management Act 2002 expired and Canterbury Earthquake Response and Recovery Act 2010 was enacted on September 14. The act established the Canterbury Earthquake Recovery Commission comprising government-appointed commissioners and local Mayors, with the responsibilities of advising the government on recovery priorities and coordinating recovery activities.

After the February 22, 2011, earthquake, the government introduced a significant change in the arrangements of recovery governance (NZ Government, 2011). The government declared NZ's first national state of emergency. New legislation was in place recognising the required powers and authorities to implement recovery policy (Brookie, 2012; Faisal, 2019). Under the new legislation, the Canterbury Earthquake Recovery Commission was disestablished, and Canterbury Earthquake Recovery Act 2011 set up a new public service department 'CERA'. NZ Government (2011) states that the CERA was formulated and established based on the lessons learned from the Canterbury's September 2010 earthquake and international experience of recovery intended to lead and coordinate the recovery.

To facilitate recovery, CERA reserved the wide power to relax, extend or suspend laws and regulations deemed necessary for earthquake recovery (NZ Government, 2011). New legislation empowered CERA to acquire land, or an interest in land, demolish buildings and rebuild or change the use of an area, compulsorily as necessary. The Greater Christchurch Group (2017, p. 8) under the Department of the Prime Minister and Cabinet reports that "the regulation-making power proved to be an effective recovery tool, pinpointing and resolving short-term or specific problems with primary legislation". The government confirmed that it was committed to retaining local democracy and community engagements. In addition to the roles in CERA, the councils would have the key roles of community advocates and authority of consent decision and service delivery. CERA was also directed to collaborate with other central government agencies in contribution to the Canterbury recovery (NZ Government, 2011). The inception of a new government department in the context of response and recovery from Canterbury earthquakes reflects the insufficient legislative arrangements and limitation of existing government agencies to deal with the large-scale disaster.

CERA transferred its responsibilities to other government agencies at the time that it ceased to exit. The expiry of Canterbury Earthquake Recovery Act 2011 in April 2016 allowed the Greater Christchurch Regeneration Act 2016 in place enabling the successive agencies (Otakaro Limited, Regenerate Christchurch and Development Christchurch Limited) to continue recovery, rebuild and regeneration (NZ Government, 2016).

Recovery of built environment in CBD

The political dimension of post-disaster recovery undermines the historic traits of the city to achieve a radical redevelopment (Amore, 2016a; Verderber, 2009). Moreover, the market-led policymaking focuses on the post-disaster lucrative economy of urban recovery (Klein, 2007; Verderber, 2009).

Based on the geotechnical information of land damage, every residential property in Christchurch was categorised into Red Zone (not recommended for redevelopment) or Green Zone (suitable for repairing and rebuilding). Over 40% of the central city residential properties were rated as 'unsafe' or 'restricted access' displacing a large number of households. CERA demolished 1,434 buildings across the city, comprising more than 50% of the building stocks in CBD (CERA, 2016a). The numbers changed in every stage of the recovery due to the continuing aftershocks and assessment of damage and insurance claims (Greater Christchurch Group, 2017).

Economic feasibility was the first consideration in deciding the fortune of heritage buildings; "the signals were that there's not much money, and there were also very strong signals coming from Government that recovery cannot be impeded by agonising over issues like heritage" (Jason Dowse, Principal Advisor, Policy, CERA, 2016a). The actively functioning historic sites until the earthquake sequence comprise Churches, Theatres, Arts Centres, Government Buildings, Hotels, Retail Premises, Office Buildings and Recreational Facilities both in public and private ownership. The cultural precinct in the city centre was one of the main tourist attractions. But the importance of heritage recovery was not acknowledged in the central government-led recovery and regeneration process; 45% of the heritage buildings in the central city were demolished by December 2011. Building Code requirement for the mandatory earthquake strengthening of the historic buildings was the turning point of post-quake heritage governance, leading to demolition or restoration of masonry buildings. The post-quake cordoned area (92 hectares) in the CBD was then the largest construction work site in Australasia until June 2013 (Amore, 2016; Greater Christchurch Group, 2017).

CCC was in the initial lead role to develop the CBD Recovery Plan. "Creating history with the central city plan [...] this is the most important time in Christchurch's history since the city was established more than 160 years ago" (Bob Parker, Mayor of Christchurch, CCC, 2011). In consultation with the community, CCC documented a redevelopment vision for the city centre. The draft central city plan in early 2012 incorporated the ideas collected from the public engagement activities attracting 10,000 people. But in the later stage of finalising the 'Christchurch Central Recovery Plan', government interventions mandated CERA as the delivery entity to lead and facilitate the central city. Christchurch Central Development Unit (CCDU) was established within CERA in April 2012 (Greater Christchurch Group, 2017). CCDU prepared a blueprint for central city rebuild, based on the initial draft of CCC to deliver a redesigned compact CBD. The blueprint remains a foundation plan for the regeneration of Christchurch, which aims to create a high-density business and retail district exceeding the pre-earthquake level. The redesigned compact CBD is bounded by Lichfield Street in the south, Avon Riven to the north and west and Manchester Street in the east (CERA, 2012).

The recovery blueprint acknowledged the city centre street patterns as the longest lasting heritage and the core architecture of Christchurch. Modified transport networks, redesigned city blocks with new lanes and country yards, transitional architecture and establishment of cultural, health, retail, civic, performing arts, sports, convention and innovation precincts re-shaped the city's built environment. Clustering the activities in the large empty blocks throughout the central city is expected to provide the catalyst for redevelopment and bring people back to the central (CERA, 2012). The proposed mixed-use zone on the edge of the central city would provide potential to live near places of works and support business and retail in the core. Legislative and regulatory changes in land use and building codes changed the urban landscape of Christchurch and subsequently transformed the economic geography.

Crisis-driven urban regeneration, tourism recovery and conference industry

The central city was in gradual decline before the earthquake, with the availability of commercial properties outstripping market demand. Moreover, city centre businesses were migrating to the new commercial areas on the city fringe, further reducing demand in the central (Greater Christchurch Group, 2017). Following the decline in the late 1990s, the CCC developed a central city revitalisation strategy in 2006 and stated

> the success of Christchurch is tied directly to the success of our Central City. The Centre's importance is borne out through its historical significance as the birthplace of our community, its current role as the commercial, cultural and social centre of the region.
>
> (CCC, 2006, p. 4)

However, the city centre was still declining in 2010.

After the earthquake, there was a clean slate to redesign and revive the city centre. Government agencies promoted the importance of the tourism sector in the city rebuild and economic recovery planning. City centre recovery blueprint frames that as a unique opportunity to update building stocks and revitalise the central city, bringing business and people back into the CBD offering accessible tourism opportunities, a vibrant city centre combining retail businesses, professional services, tourism and hospitality.

The shutdown, long-term cordon and staged reopening has meant to a continued dislocation of city centre businesses, workers, residents, students and visitors. The critical challenges in the post-quake tourism sector were to fight back with the devastating impacts, managing the international media attention, lack of infrastructure, decreased number of visitors, changes in visitor demography and finally the place (re)making. The Christchurch International Airport chief executive described the situation as the biggest crisis of the Christchurch tourism sector and urged for collaborative efforts to ensure its presence on the tourist map (*Otago Daily Times*, 2011, April

18). The CCT marketing team adopted the strategy of telling positive stories to consumers, media and travel sellers. The collaborative approaches of the government agencies, CCT and Christchurch international airport recognised the importance of tourism in city's economy and jointly injected multi-million dollars to rebuild the tourism activities, remarket the destination and rebound quickly (*Otago Daily Times*, 2011). Christchurch mayor Bob Parker (2011) stated in an interview with *Otago Daily Times* (2011, April 18) that "we have to get out now with the message that our tourism industry is still operating and that Christchurch is the gateway to the South Island, which is open for business and as beautiful as ever". CCT highly promoted the remaining attractions, nature-based activities, cultural diversities, heritage and innovative ideas that emerged in the quake's aftermath.

"Functioning businesses are critical to economic and community recovery in many ways" (Resilient NZ, 2015). The recovery of the retail sector in CBD began with the relocatable container structure 'Re:START' mall in October 2011, an innovative initiative of the central city retailers in partnership with both the public and private institutions. The government allowed the use of a part of the cordoned Red Zone, resulting in an iconic venture of the business community. Adaptability and entrepreneurial strategy of a group of central city businesses provided a catalyst for further redevelopments in the surrounding area. The success of the project re-established the area as a prime retail destination and emerged as a unique landmark and famous tourist attractions, significantly contributing to the central city tourism recovery. The concept of the 'Re:START' container mall in the heart of the devastated city centre was one of the unique selling propositions of the Christchurch destination marketing campaign, highlighting the diversity, resilience and opportunities in the transitional city spaces (CCT, 2011).

The city's ability to host conferences has been significantly compromised by the closure of venues and hotel conference centres. Uncertainty in re-opening timeframes and demolitions resulted in a 43% reduction in conference capacity. All the scheduled large conferences in Christchurch were cancelled or relocated to other competitor cities like Auckland, Wellington and Queenstown. CCB had a radical shift in the transitional strategy formulation; it downsized its professional team and focused on promoting a very different range of products like one-day conferences and smaller multi-day conferences in line with the carrying capacity of the existing infrastructure (CCT, 2011). In 2012, Christchurch hosted only 2% of the conference delegate days in NZ, primarily generated by the conferences of 100–250 delegates (CCT, 2016).

Auckland, Wellington and Rotorua have been very active in capturing Christchurch's share of the international conference market since Christchurch's exit in 2011. As the city continued to recover and supply of conference facilities (mid-size venues and hotel rooms) increased, the CCB promoted a one-day exhibition in September 2013 highlighting their 56 members, the first trade exhibition showcasing the venues, accommodation and services available for hosting conferences in post-quake Christchurch (Wood, 2013). CCB continued to focus on the domestic market and re-entered the key target market Australia in 2015 encountering the negative impressions of Christchurch. Professional networking, updating events, increased bidding activities and collaborations with other tourism management agencies were the primary tools to re-engage with the international conference market. Christchurch's share of national conference industry increased to 9% in 2016. CCB manager Caroline Blanchfield signifies the growth of the 'business tourism' and notes that international convention delegates are spending twice as much as other international visitors (McDonald & Small, 2017). Christchurch's national contribution to the tourism sector recovered to 7.8% in 2016. Christchurch's conference industry is also rebounding, but remains on slow growth and plans the next step up on the opening of the new convention centre precinct to be completed in early 2020 (Wood, 2015a; McDonald, 2016a).

The trade associations, local government and regional tourism organisation argue that the convention centre was the engine room for the city's tourism industry. "Tourism will play a key role in restoring economic prosperity to Christchurch's Central City.... a world-class convention centre in the heart of the Central City is critical for Christchurch's economic recovery" (CCC, 2011, p. 133). CCT chief executive Tim Hunter explains the necessity of a rebuilt convention centre "we know from talking to people right across the tourism sector that it is absolutely vital to reinstate a centre" (Sachdeva, 2011, August 11). The delivery of a new convention centre is delayed due to the critical issues of site selection, design and the public-private cost sharing negotiation. Sachdeva compiled the responses of industry leaders representing Canterbury Employers' Chamber of Commerce, CCT and Central City Business Association in the context of City Council's demolition decision of the damaged convention centre. He reported that the quick rebuild of the convention centre is vital for the tourism sector to retain its share in the highly competitive market of the conference industry. Bruce Garrett, the chairman of the Tourism Industry Association's hotel section, noted the frustration of the hotel industry in complexities and delayed delivery of the tourism convention centre. In this vein, the NZ Property Council confirmed that the delays costed the city millions of dollars in capital flights (Sachdeva, 2011, August 11).

Government finally announced the public funding to cover the full cost of the Christchurch Convention Centre, set to be NZ's second largest convention facility for 200–2000 delegates (Wood, 2015, March 25; Tourism Australia & Fleming, 2017, September 25), with the primary aim to minimise uncertainty in the Christchurch tourism industry and to catalyse hotel development. Christchurch Regeneration Minister Nicky Wagner disclosed that the full government funded convention centre is expected to bring direct benefit to Canterbury, NZ$321 million in the first eight years and

NZ$ 57 million to NZ$ 60 million every year after that (McDonald, 2017, August 23). Canterbury Employers' Chamber of Commerce General Manager Leeann Watson symbolised the convention centre as a vital piece of making an important statement to instil confidence of the city businesses (McDonald, 2016, June 30).

As a part of the tourism-led economic policy, the government allocated NZ$ 34 million over four years to bring business events to NZ. The fund assisted CCB in the bidding process. Recent successful bids had been for five international professional conferences in 2015–2020, estimated to host thousands of international visitors in Christchurch (Wood, 2015a). The noteworthy strategy of CCB was engaging with the Christchurch-based leading professionals in the respective fields of the conferences. CCB manager in her press interview with Wood (2015a) mentioned the help of Christchurch professionals in winning the bid of international conferences such as the Asia Oceania Society of Physical and Rehabilitation Medicine, the World Congress of Endometriosis and the Annual Meeting of the Pacific Association of Paediatric Surgeons.

Quake-shattered Christchurch witnessed a significant shift in visitor demography.

> There's still a bit of reluctance to come and visit Christchurch....
> There's still however a feeling from some of the wholesale trade that Christchurch doesn't justify quite as long a stay as it did pre-quake, they're waiting for the city to be rebuilt.
>
> (Bruce Garrett, The Tourism Industry Association's Canterbury hotel sector chairman in McDonald, 2016, February 24)

In the 12 months to April 2017, the number of Chinese visitors in Christchurch remarkably increased from April 2010 (from 1.3% to 8.5% of the international visitor arrivals), but the Australian arrival didn't return to pre-earthquake level (fell from 57.6% to 47.3% of the international visitor arrivals). Visitors spent 2.7 million nights (81.8% in March 2010) in commercial accommodation in greater Christchurch. The central city accommodation capacity reached 2,042 units (but 61% less than the available units in March 2010), and visitors spent 753,713 guest nights – nearly 55.5% less than the March 2010 level (*NZ Herald*, 2017, June 11).

The recovery process and multiplier effects of rebuild dollars tended to amplify regional trends against national trends of business revenue growth in key sectors. As the rebuild gathers momentum, revenue trends (2011–2015) in Christ Church showed a strong growth to 120% in the construction sector, compared to 45% nationally. Following a revenue decline of 17.3% (compared to a national gain of 11.9%) between 2011 and 2013, the accommodation and food services were on the rebound with a 10% growth compared to the national growth of 2% in 2015 (KPMG, 2017).

Conclusion

As NZ$ 100 million are being spent on the rebuild every week (NZ Trade and Enterprises, 2016), a new city is now taking shape. "The $40 billion post-earthquake rebuild continues at a pace to deliver an attractive, modern, resilient and future-proofed city" (Canterbury Development Corporation (CDC), 2015, p. 2). The Canterbury Earthquake Recovery Act 2011 overruled the existing legislation and enabled the government to acquire the designated land for the anchor projects to achieve the development of the Christchurch city centre in 'depoliticised climate'. The redevelopment blueprint was justified through the benchmarking of international best practices to attract tourists and development capital (Amore & Hall, 2016). However, the lack of private investments for anchor projects and complexities in recovery process reframed the highly market-driven approaches of urban regeneration. The long-term cordon in the city centre, demolition of the built heritage, political interventions, transitional architecture, anchor projects and city precincts altered the city spaces and identity.

The aftermath of the Christchurch earthquakes demonstrates the complex issues of recovery from the devastating natural disaster: how the recovery momentum influences the tourism sector and conference portfolios. The earthquakes caused the temporary extinction of the partial population of conference events in Christchurch, resulting in a significant disruption in the tourism economy. Due to resource constraints, the portfolio of large multi-day international conferences was lost to competing destinations. As a strategic business unit of CCT, CCB adopted collaborative approaches (with government agencies, regional tourism organisations, professional conference organisers and conference service providers) and focused on the portfolios of one-day small and medium conferences. But Christchurch's conference industry redeveloped a large conference portfolio as conference facilities continued to recover.

Post-disaster evolution of a revived conference industry is a complex interplay of the intrinsic and extrinsic factors pertaining to the local, regional and national economies. The redesign and redevelopment of the built environment, public and private investment in tourism infrastructure and regeneration strategy for the affected region emerged as key aspects in shaping the trajectory of a post-disaster conference industry. Stakeholder agility is found to be the decisive factor of event population management in an adverse environment. As no destination is immune to disasters (Faulkner, 2001), strategic (re)alignment during a self-sustaining recovery is indispensable for a conference industry to survive and thrive in times of environmental uncertainty.

References

Amore, A. (2016). The governance of built heritage in the post-earthquake Christchurch CBD. In C. M. Hall, S. Malinen, R. Vosslamber, & R. Wordsworth

(Eds.), *Business and Post-Disaster Management: Business, Organizational and Consumer Resilience and the Christchurch Earthquake* (pp. 200–230). Oxon and New York: Routledge.

Amore, A., & Hall, C. M. (2016). 'Regeneration is the focus now': Anchor projects and delivering a new CBD for Christchurch. In C. M. Hall, S. Malinen, R. Vosslamber, & R. Wordsworth (Eds.), *Business and Post-Disaster Management: Business, Organisational and Consumer Resilience and Christchurch Earthquakes* (pp. 181–199). New York: Routledge.

Anderssson, T. D., & Getz, D. (2009). Tourism as a mixed industry: Differences between private, public and not-for-profit festivals. *Tourism Management, 30*, 847–856.

Ashworth, G., & Page, S. J. (2011). Urban tourism research: Recent progress and current paradoxes. *Tourism Management, 32*, 1–15.

Baum, J. (1996). Organizational ecology. In S. Clegg, C. Hardy, & W. Nord (Eds.), *Handbook of Organizational Study* (pp. 77–115). London: Sage.

Bradley, G. (2013, August 28). Christchurch gearing up for tourism rebirth. *NZ Herald*. Retrieved from http://www.nzherald.co.nz/business/news/article.cfm?c_id=3&objectid=11115263

Brookie, R. (2012). *Governing the Recovery from the Canterbury Earthquakes 2010–11: The Debate over Institutional Design*. Institute for Governance and Policy Studies, Victoria University of Wellington.

Canterbury Development Corporation. (2015). *Christchurch - The City of Opportunity for Businesses, Entrepreneurs and Investors*. Christchurch: CDC.

Canterbury Earthquake Recovery Authority. (2012a). *Christchurch Central Recovery Plan Summary*. Christchurch: Canterbury Earthquake Recovery Authority.

Canterbury Earthqauke Recovery Authority. (2012b). *Christchurch Central Recovery Plan Summary*. Christchurch: Canterbury Earthquake Recovery Authority

Canterbury Earthquake Recovery Authority (CERA). (2014). *Canterbury Earthquake Recovery Authority Statement of Intent 2014–2018*. Christchurch: Canterbury Earthquake Recovery Authority. Retrieved from www.cera.govt.nz.

Canterbury Earthquake Recovery Authority (CERA). (2015). *Greater Christchurch Earthquake Recovery: Transition to Regeneration*. Christchurch: Canterbury Earthquake Recovery Authority.

Canterbury Earthquake Recovery Authority. (2016a). *Demolitions and Operations: An Overview of the CERA Operations Team*. New Zealand: EQ Recovery Learning.

Canterbury Earthquake Recovery Authority. (2016b). *Funding the Recovery: The CERA Perspective*. New Zealand: EQ Recovery Learning.

Canterbury Earthquake Recovery Authority (CERA). (2016c). *Recovering Christchurch's Central City: A Narrative of the First Four Years 2011–2015*. EQ Recovery Learning.

Chalip, L. (2006). Towards social leverage of sport events. *Journal of Sport & Tourism, 11*, 109–127.

Christchurch and Canterbury Tourism. (2011). *2011 Annual Report to the Trustees of Destination Christchurch*. Christchurch: CCT.

Christchurch and Canterbury Tourism. (2013). *2013 Annual Report to the Trustees of Destination Christchurch Canterbury New Zealand Trust*. Christchurch: CCT.

Christchurch and Canterbury Tourism. (2016). *Annual Report for the Year Ended 30 June 2016*. Retrieved from https://www.christchurchnz.com/media/2556/-christchurch-and-canterbury-tourism-annual-report-2016.pdf

Christchurch City Council. (2006). *Central City Revitalisation Strategy Stage 2.* Christchurch: Christchurch City Council.

Christchurch City Council. (2011). *Central City Plan.* Christchurch: Christchurch City Council.

Faisal, A. (2019). *Entrepreneurial Responses to Disruptive Transformations: An Evolutionary Perspective on the Hospitality Industry of Post-Quake Christchurch, New Zealand* (Thesis, Doctor of Philosophy). University of Otago, New Zealand. Retrieved from http://hdl.handle.net/10523/9183

Faisal, A., Albrecht, J., & Coetzee, W. J. L. (2017). The challenges of (re)gaining a sustainable business proposition post-disaster. In C. Lee, S. Filep, J. N. Albrecht, & W. J. L. Coetzee (Eds.), *Proceedings of the 27th Council for Australasian Tourism and Hospitality Education(CAUTHE) Annual Conference* (pp. 597–601). Dunedin, NZ: Department of Tourism, University of Otago.

Faulkner, B. (2001). Towards a framework for tourism disaster management. *Tourism Management, 22*(2), 135–147.

Getz, D. (2008). Event tourism: Definition, evolution, and research. *Tourism Management, 29,* 403–428.

Getz, D. (2017). Developing a framework for sustainable cities. *Event Management, 21,* 575–591.

Greater Christchurch Group. (2017). *Whole of Government Report: Lessons from the Canterbury Earthquake Sequence.* Christchurch: Greater Christchurch Group, Department of the Prime Minister and Cabinet.

Hall, C. M., Prayag, G., & Amore, A. (2018). *Tourism and Resilience: Individual, Organisational and Destination Perspectives.* Bristol; Blue Ridge Summit, PA: Channel View Publications.

Hannan, M., Polos, L., & Carroll, G. (2007). *Logics of Organization Theory: Audiences, Code, and Ecologies.* Princeton: University Press.

Henderson, J. C. (2007). *Managing Tourism Crises.* (1st ed.). Burlington: Taylor and Francis.

Hiller, H. (2000). Mega-events, urban boosterism and growth strategies: An analysis of the objectives and legitimations of the Cape Town 2004 Olympic bid. *International Journal of Urban and Regional Research, 24*(2), 439–458.

Huan, T. C., Tsai, C. F., & Shelby, L. B. (2006). Impacts of No-Escape Natural Disaster on Tourism: A Case Study in Taiwan. *Advances in Hospitality and Leisure, 2,* 91–106.

Ingham, J., & Griffith, M. (2010). Performance of unreinforced masonry buildings during the 2010 Darfield (Christchurch, NZ) Earthquake. *Australian Journal of Structural Engineering, 11*(3), 207–224.

Karski, A. (1990). Urban tourism: A key to urban regeneration. *The Planner, 76*(13), 15–17.

Klein, N. (2007). *The Shock Doctrine: The Rise of Disaster Capitalism.* New York: Metropolitan Books.

KPMG. (2017). *The Impact of the Canterbury Earthquakes on Privately Owned Businesses.* Christchurch: CDC.

Maitland, R. (2009). Introduction: National capitals and city tourism. In R. Maitland & B. W. Ritchie (Eds.), *City Tourism: National Capital Perspectives* (pp. 1–13). Wallingford; Cambridge, MA: CABI.

McDonald, L. (2016a, February 24). Canterbury tourism back to pre-earthquake levels, but city skipped. *Stuff.* Retrieved from https://www.stuff.co.nz/the-press/

business/the-rebuild/77181585/canterbury-tourism-back-to-preearthquake-levels-but-city-skipped

McDonald, L. (2016b, June 30). Surge of new hotels expected on back of Christchurch Convention Centre plan. *Stuff.* Retrieved from https://www.stuff.co.nz/the-press/business/the-rebuild/81628140/Surge-of-new-hotels-expected-on-back-of-Christchurch-Convention-Centre-plan

McDonald, L., & Small, J. (2017, August 23). Christchurch convention centre cost updated to $475m. *The Press.* Retrieved from https://www.stuff.co.nz/the-press/news/96090557/convention-centre-cost-updated-to-475m

NZ Government. (2011). *Canterbury Earthquake Response and Recovery Act 2010.* Wellington: NZ Government.

NZ Government. (2016). *Greater Christchurch Regeneration Act 2016.* Wellington: NZ Government.

NZ Herald. (2017, June 11). Christchurch 'not a broken place', tourists returning steadily. Retrieved from http://www.nzherald.co.nz/nz/news/article.cfm?c_id=1&objectid=11874158

NZ Trade and Enterprises. (2016). *Showcasing Canterbury's Collaborative Innovation.* New Zealand: NZ Trade and Enterprises.

Orchiston, C. (2012). Seismic risk scenario planning and sustainable tourism management: Christchurch and the Alpine Fault zone, South Island, NZ. *Journal of Sustainable Tourism, 20*(1), 59–79.

Otago Daily Times. (2011, April 18). New venture to boost Christchurch tourism. Retrieved from https://www.odt.co.nz/news/national/new-venture-boost-christchurch-tourism

Patterson, I., & Getz, D. (2013). At the nexus of leisure and event studies. *Event Management, 17*, 227–240.

Pforr, C. (2009). Crisis management in tourism: A review of the emergent literature. In C. Pforr & P. Hoise (Eds.), *Crisis Management in the Tourism Industry* (pp. 37–52). Surrey and Burlington: Ashgate Publishing.

Resilient NZ. (2015). *Contributing More: Improving the Role of Business in Recovery.* Christchurch: Resilient NZ.

Ritchie, B. W., & Campiranon, K. (2015). Major themes and perspective. In B. W. Ritchie & K. Campiranon (Eds.), *Tourism Crisis and Disaster Management in the Asia-Pacific.* Wallingford, OX: CABI.

Rittichainuwat, B., & Mair, J. (2012). Consumer motivations for attending travel shows. *Tourism Management, 33*(5), 1236–1244.

Rogers, T. (2013). *Conferences and Conventions 3rd edition A Global Industry.* New York: Taylor and Francis.

Sachdeva, S. (2011, August 11). Quick rebuild 'vital' for tourism. *Stuff.* Retrieved from http://www.stuff.co.nz/the-press/news/christchurch-earthquake-2011/5425694/Quick-rebuild-vital-for-tourism

Spirou, C. (2011). *Urban Tourism and Urban Change: Cities in Global Economy.* New York: Routledge.

Stevenson, J. R., Seville, E., Kachali, H., Vargo, J., & Whitman, Z. (2011). *Post-disaster Organisational Recovery in a Central Business District Context: The 2010 & 2011 Canterbury Earthquakes.* Christchurch: Resilient Organisations.

Swiss Re. (2017). Swiss Re Financial Report 2016. Retrieved from http://reports.swissre.com/2016/financial-report/responsibility/natural-catastrophes-and-climate-change.html

Tourism Australia, & Fleming, G. (2017, September 25). A city of opportunity. Retrieved from http://www.nzherald.co.nz/business-travel/news/article.cfm?c_id =813&objectid=11924982

Tourism Ticker. (2017, June 12). Christchurch back on tourist map. Retrieved from http://tourismticker.com/2017/06/12/christchurch-back-on-the-tourist-map/

Verderber, S. (2009). The unbuilding of historic neighbourhoods in Post-Katrina New Orleans. *Journal of Urban Design, 14*(3), 257–277.

Wood, A. (2013, May 27). Eager tourism industry keeps eye on conferences. *Stuff.* Retrieved from http://www.stuff.co.nz/business/industries/8719209/Eager-tourism -industry-keeps-eye-on-conferences

Wood, A. (2015a, March 25). City claws back lost conference trade bit by bit. *Stuff.* Retrieved from https://www.stuff.co.nz/business/industries/67496264/city-claws-back-lost-conference-trade-bit-by-bit

Wood, A. (2015b, May 20). Prime Minister says Christchurch convention centre costs "prohibitive". *Stuff.* Retrieved from https://www.stuff.co.nz/the-press/ business/68697238/Prime-Minister-says-Christchurch-convention-centre-costs-prohibitive

9 A critical view on mega-events economic impacts studies

Milan 2015 and beyond

Jérôme Massiani

Introduction

Mega-events, such as the Olympics or international exhibitions, are often perceived as a real opportunity for image, urban renewal and economic development. Focusing on this latest aspect, Expo 2015 is no exception with expectations as high as 10–30 billion euros of increased added value (Airoldi *et al.*, 2010; Dell'Acqua, Morri and Quaini, 2013). Focusing on the sole impact of event visitors, the same impact studies expected 29 million visits (20 million visitors) and able to generate 4.3–4.8 billion euros of added value. Such quantifications however appear discussible as they do not really consider whether visitor expenditures are additional or substitutive of others. Rigorously, a share of expenditures made for the Expo cannot be considered additional for the investigated territory, whether Italy as a whole, whether Lombardy, the host region, alone. For instance, some foreigners would have come to Italy even without the event. Similarly, expenditures of the local population are, at least partially, substitutive of other expenditures on the national territory.

The purpose of this chapter is to evaluate the impact of Expo visitors on the Italian (and regional) economy considering only the additional share of these expenditures. To do so, one needs to consider a non-observable counterfactual situation, conceptually: what visitors would have done without the Expo. In this context, various methods can be proposed like econometric studies of consumers' expenditures, Differences in Differences, and others. These methods have their benefits but, in many cases, the necessary data are not available. This is the case for Milano 2015 where Italian national household surveys underwent restructuring in 2013, too close to 2015 to use it without a high degree of risk. This chapter will implement instead a questionnaire-based method to estimate the actual impact of Expo visitors spending. While questionnaires are imperfect measures of counterfactual behaviour, they can be used to provide an order of magnitude of the investigated impacts.

The chapter runs as follows: Section "Substitution effects in the economic analysis of mega-events" examines how the literature analyses the substitution effect. Section "Expo 2015 survey" presents our survey instrument. Section "The impact of Expo 2015: a sizeable correction" measures the impact of

Expo visitors. This is done both for Italy as a whole and for Lombardy alone. Section "Conclusions" discusses the results and draws conclusions.

Substitution effects in the economic analysis of mega-events

The international literature underlines that many quantifications of the impact of mega-events can be distorted if no careful attention is dedicated to some effects, sometimes taken as synonym, sometimes not, like substitution, displacement or deadweight (Felsenstein and Fleischer, 2003; Vanhove, 2005; Oxford Economics, 2012). More in detail, the debated issues relate to the various corrections that are necessary to correct the distorted Naïve Input-Output calculation. First, local expenditures cannot be entirely considered as additional for the investigated territory (Tyrrell and Johnston, 2001). Second, foreign expenditure is also at risk of overestimation. Visitors that had planned to visit the area even without the event, but change the dates of their trips, should not be accounted for (Getz, 1994; Ryan, 1998). This is the same thing for visitors who would have come anyway and for foreigners that would be present in the area without the event but decide to attend it (Crompton, Lee and Shuster, 2001). One should also consider crowding out: reduction in ordinary tourism when the event takes place (Baade, Baumann and Matheson, 2008). To consider these types of behavioural responses Preuss' terminology is used (Preuss, 2005), which identifies various behaviours, for instance stayers, people who decide to stay in their residing city for the event, or time shifters, people who change the period of their visit to a city in order to attend an event.

One problem is that the terminology in use among researchers is not homogeneous. Various scholars use various expressions like "substitution effect", "displacement", "crowding-out" with a different meaning. Notwithstanding this difficulty, the quoted literature agrees on the need to exclude, at least in part, local visitors' expenditures, because a sizeable share of this expenditure is substitutive of other expenditures in the territory.

In order to represent correctly these different behavioural responses, several methods are available.

A first approach is a correction *ex abrupto* by elimination of the local population (Webber, White and Smith, 2011). For instance the French *Ministère de la Culture* (Nicolas, 2007) recommends to compute benefits after subtraction of the expenditures of local visitors. However, this method entails two approximations. The first one is that some local visitors may be additional (for instance, stayers: they would have left the territory without the event). The second one is that foreign visitors may not be additional (time switchers: they would have come in the territory in a different period). The former effect would increase the impact compared with a computation based on the exclusion of locals, while the latter would reduce it. On the whole, the exclusion of the local population may lead to an over- or an under-estimate. However evidence suggests that there is more risk of overestimation because the non-additional component of foreigners is larger than the additional component of the local population. Generally, the risk of distortion makes it necessary to consider alternative, more consistent, methods.

Other possible approaches may rely on:

- Econometric analysis of a number of events. This analysis examines the evolution of expenditures on a number of territories that underwent an event "shock"
- Econometric analysis of household expenditure surveys in a single territory that hosted an event
- Questionnaires among visitors, asking consumers what they would have done if the event had not taken place.

The first method, econometric analysis, has the advantage of consolidating the findings on a potentially large number of cases. However, it has the limitation that each event can be thought as irreproducible and this casts doubt on the real validity or on the face value of these findings when applied to a specific event.

The second method, household expenditure surveys, is attractive and would allow us to estimate how a given event had an impact on the consumption of local population. Obviously, this would however leave aside the question of foreign visitors. Another issue is that national household surveys are sometimes restructured (the format of the data collection is changed), which creates structural breaks in time series and hinders the validity of time series analysis. For instance, in the case of Milan 2015, the Italian national household survey underwent a strong restructuring in 2013, thus making it difficult to use the corresponding time series around that year. In a parallel work, we undertake a similar analysis for the impact of Torino 2006 Olympics, as 2006 was not impacted by this data issue. Temporary results indicate that the change in consumption of the local population is much lower than any possible quantification of the event-related expenditures of the same population.

The third method, based on questionnaire, has the obvious limitation that it relies on declaration from visitors. However, it has the advantage of covering a large set of populations (namely visiting foreigners and visiting locals). This approach is also beneficial in that it allows us to unveil the ingenuity of many IO-based calculations that erroneously consider all event-related expenditure as additional. There is now a large number of academic papers on this issue (McHone and Rungeling, 2000; Crompton, Lee and Shuster, 2001; Felsenstein and Fleischer, 2003; Preuss, 2005; Barget and Gouguet, 2007, 2010, 2011; Solberg and Preuss, 2007; BOP, 2011; Taks *et al.*, 2011)

Based on this approach, a survey is directed to visitors for Expo 2015 to measure the additional versus substitutive component of their expenditures. In the next sections, we present the method and results of a survey performed to account only for the additional impact.

Expo 2015 survey

In this section, we first provide basic elements on the event and subsequently present the survey instrument.

Expo Milano 2015

The event took place from May 1 to October 31, 2015. An ex-ante prediction forecasted 29 million visits and 520 million ticketing revenues (Comitato di candidatura, 2006). Ex post outcomes were significantly reduced with a number of visits around 22 million and net ticketing revenues of 427 million euro (Expo 2015 SpA, 2016). The economic impact of Expo visitors was estimated ex-ante at 4.3–4.8 billion euro of increase in the national added value (CERTeT, 2010; Dell'Acqua, Morri and Quaini, 2013). These figures were part of a larger study that dealt with various items such as illustrated in Figure 9.1.

The survey

In this section, we present the questionnaire and how it was operated. We selected the question phrasing used in the study of Edinburgh Festival impacted by BOP consulting (BOP, 2011). This choice was motivated by the clarity and simplicity of expression, the consolidated experience of the authors with this formulation and the possibility to easily compare our results with other ones. The key questions relate to the activity to which visitors renounced to attend the event:

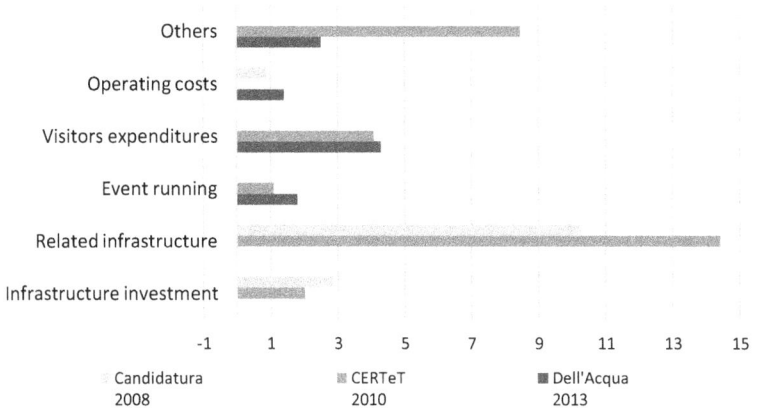

Figure 9.1 Various impacts considered in Milan 2015 impact studies (bln euro).

If today you would not have come to Expo, what would you probably have done?

- I would have stayed home or at work
- I would have done something else in Lombardy
- I would have done something else in another region of Italy
- I would have stayed abroad.

Apart from these, other questions, some for double checking, were also asked to visitors. Some questions also related to the alternative expenditure of money spent for visiting the expo.

The survey was answered by 873 visitors, in five languages in each of the four accesses to the Expo area during 12 days distributed between July 21 and October 24, 2015.

Two technical questions should be touched upon:

First, the question of "representativity" is often discussed about surveys; this point may deserve clarification. For instance, Schouten and his co-authors indicate "The concept of representativity is often used in survey research, but usually it is not clear what it means" (Bethlehem, Schouten and Cobben, 2009). Based on these works, Kruskal and Mosteller show that representativity has several meanings:

> (1) general acclaim for data, (2) absence of selective forces, (3) miniature of the population, (4) typical or ideal case(s), (5) coverage of the population, (6) a vague term, to be made precise, (7) representative sampling as a specific sampling method, (8) as permitting good estimation, or (9) good enough for a particular purpose.
>
> (Kruskal and Mosteller, 1979, 1980)

Based on this observation, it seems appropriate to concentrate on more univocal concepts like equiprobability and sample cardinality.

Second, we should consider equiprobability. The survey process aimed at covering the various visitors' profiles. Each survey day, some entrances gates were selected randomly and all visitors in these queues were interviewed. An interesting aspect was that only very few visitors refused to answer the questionnaire; certain days, no refusal was observed. Thus, the data collection took place under much more favourable conditions than is usually the case. This increases the quality of data and especially reduces the risk of auto-selection.

The impact of Expo 2015: a sizeable correction

Based on these data, one can better know the profile of visitors and which expenditure is actually additional.

On the first point, the local visitors appear strongly dominating. This applies both at the national scale: 85.4% of visitors are Italian, and at the regional level: 37.5% of the visitors are from Lombardy, 8.8% from the neighbouring region of Emilia-Romagna and 7.9% of Piedmont. This already provides an indication on how the real (additional) impact of the event may be drastically resized when properly treating the domestic demand. It also becomes possible to compute the actual share of additional visitors with proper exclusion of visitors who would have come/stayed in the investigated territory in every case.

At this stage, it is not too early to define what is meant by "local". For the specific purpose of our analysis, we will consider two scales: a national scale and a regional scale (Lombardy). These two levels of analyses also allow us to shed light on the regional redistribution effect of mega-events. Expo 2015 triggered a major flow of resources from other Italian regions to Lombardy: visitors from say, Lazio, spend money in Lombardy rather than in another region.

The impact for Italy

Table 9.1 summarizes the information on origin and alternative activity. It shows that only 1% among 85% of Italian visitors are to be accounted as additional. Similarly, only 7.6% among 15% of foreign visitors are to be considered as additional. This has strong implications on the estimation of the impact. On the whole, little more than 8% of the visitors are actually additional at the national scale.

Table 9.1 Quantification of additional visitors for Italy

Origin		What would I have done if I would not be in the Expo today				% for each origin		% on visitor's total*	
Area	%	Lombardy	Other regions	At home	Abroad	Substi-tutive	Addi-tional	Substi-tutive	Addi-tional
Abroad	15%	31.5%	16.5%	29.1%	22.8%	48%	52%	7%	7.6%
Lombardy	37%	25.1%	1.8%	72.2%	0.6%	99.1%	0.6%	37.1%	0.2%
Rest of Italy	48%	12.9%	31.8%	54.1%	0.9%	98.8%	0.9%	47.4%	0.5%
								91.5%	8.3%

Source: Our elaboration.
Reading of the table: 72% of visitors from Lombardy state that without the Expo they would have stayed home. These and two other categories should be considered substitutive for Italy. On the whole, these three categories of Lombard respondents represent 99% of the Lombard visitors and 37% of the total visitors.

Legend:

X%	% Additional
Y%	% Substitutive

Visitors spending

Based on these data one can estimate the impact of visitors' expenditures. In doing this, we operate successive definitions of what an "additional" expenditure can be. Actually, the information in Table 9.1 does not cover all possible situations. One should also consider the possibility that the event also had an impact on the length of stay: a visitor may not be additional, but may have increased the length of his stays. In addition, one can also consider or at least discuss the speculative assumption that the local consumers increased their consumption in substitution to savings. To do so, we need data on daily expenditures of various visitors. We rely on data provided by CERTeT, a research centre specialized on tourism, whose data are presented in the appendix.

Additional visitors can be foreigners (that would not have come to Italy without the event) or Italians (that would have gone abroad without the event).

Non-additional foreigners may have an impact through two mechanisms:

1 An increase of daily consumption compared with no event situation
2 A longer stay.

The first aspect is already considered in the expenditure assumptions, at least if it has been properly quantified. Thus, no correction is necessary, at least for conceptual consistency.

The second aspect can be taken in consideration based on the survey data about extended stays of the non-additional visitors. This results in extra expenditure of 0.33 billion euro (see Box 9.1 for details).

BOX 9.1 THE EXPENDITURE OF FOREIGNERS

The expenditure can be expressed as follows:

Additional expenditures of foreigners: $N \times \sigma_s \times \alpha_s \times g_s \times S_s$
Additional expenditure of Italians: $N \times \sigma_i \times \alpha_i \times E_i$
Additional expenditure due to longer stays from non-additional visitors: $N \times \sigma_s \times (1 - \alpha_s) \times g'_s \times S_s$

With

N   total visitors (20 million, based on 21.5 million after subtraction of workers, volunteers and others)
σ_s fraction of foreigners (our survey)
α_s fraction of additional among foreigners (our survey)
g_s average stay duration for each foreigner (1.4 days/visit, our survey)
S_s daily expenditure of foreigners (€171/day, data from CERTeT)
σ_i share of Italians (85%, our survey)
α_i fraction of additional Italians (our survey)
E_i alternative expenditure for additional Italians (980 € alternative expenditure abroad)[1]
g'_s additional length of stays for non-additional foreigners (our survey)

1 Based on Confesercenti, ("Estate 2015, Confesercenti, Turismo in ripartenza. In vacanza 32 milioni di italiani, quasi 2 milioni in più del 2014", 5 giugno 2015). Our estimate is a bit larger considering expenditure for travelling abroad.

A hypothetical increase in consumption

A more generous, although speculative, assumption is to consider that local population partly increased consumption, rather than merely substituted it. The point here is not that the event increased the final consumption of local population. It normally did, thanks to the indirect and induced effects of the foreign tourist. The point rather deals with the consumption patterns, such as consumption propensity, that may alter the impact of the event.

To measure this tentatively, we included other questions in the survey. This point is, we reckon, speculative. Actually, to assess its realism, we also used a series of checking questions. It appears that people are sometimes inconsistent in answering questions about the alternative use of money: interviewees sometimes respond incoherently depending on the question phrasing. It seems harder for people to answer a question on the alternative use of money ("how would you have spent your money...") rather than the alternative use of time ("what would you have done ..."). One of the possible reasons is that the alternative use of a given budget can relate to another period of time. While the question on what the person would have done in a given day refers, by definition, to the same period. More fundamentally, budgets are not earmarked, which makes the notion of "event expenditure" an elusive one. Aware of these difficulties, we posit that the most scientific approach is to report our findings while expressing necessary caveat on their validity.

The outcome of this analysis is provided in Figure 9.2. This figure represents how the increase in added value changes when a larger share of the local population's expenditure is actually additional, a ratio we label as

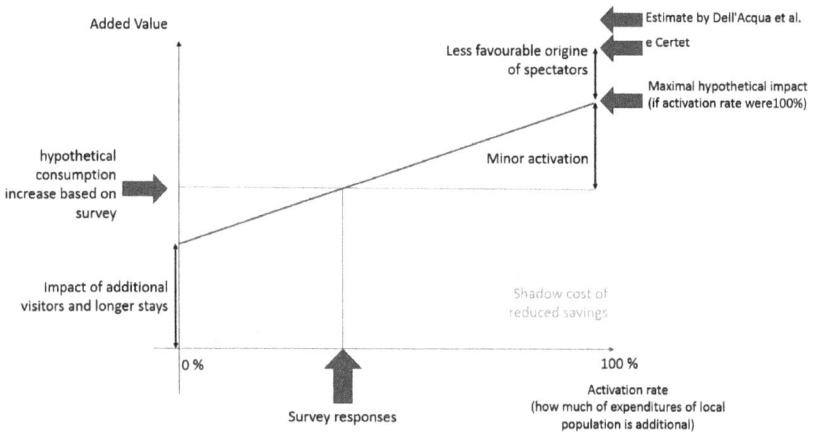

Figure 9.2 Activation of local expenditure: the spectrum of possible impact magnitudes.

"activation rate". Interestingly, even in the extreme assumption of a 100% activation rate, the expenditure created by the event would generate a smaller added value than the one assumed by ex-ante impact studies. One reason for this smaller impact is that the share of visitors coming from abroad and south Italy was smaller than expected, while these two categories have on average a higher expenditure per each visit to the Expo.

It would be relevant to search for evidence to validate the assumption of net increase of consumption propensity of local population. Some econometric results exist but they are unconvincing or provided with a low level of documentation. We illustrate this with a long quote of Oxford Economics:

> To test for a happiness effect from hosting a major event on consumer spending an additional variable has been inserted into the equations that explain consumer spending in Oxford Economics' Global Macroeconomic Model. The variable takes a value of one when the major event occurs and zero at all other times. If it is found to be positive and statistically significant for host countries it is consistent with the idea of there being a stimulus to consumption from the happiness effects associated with being host. However, while it may be consistent with this premise, it does not prove this is the case as the increase in foreign residents entering the country for the events is likely to boost consumer spending. The analysis was carried out for the three types of major events in Europe since 1980. The results are decidedly mixed, being consistent with a happiness impact from hosting an event on consumer spending for some major events but not for others.
>
> (Oxford Economics, 2012)

Some local stake-holders' testimonies also support the idea of strong substitution effects. Table 9.2 summarizes the various possible assumptions and the arguments that support them.

Table 9.2 An increase in consumption: an overview of possible assumptions

0% additional	*Partly additional*	*100% additional*
• Assumption from various CGE (0) • Consumption reduction in results of CGE (0) • Econometric models (Oxford Economics, 2012) (+)	• Our survey results: "the money I have spent for Expo, I would have spent on other things in any case" (+). • Milan shop owners' statements[a] (+). • Consumption monitoring (to be made)	• There is no proof of the contrary" (0)

[a] Complaints of local shop holders as reported in economic journals: Il Sole 24 Ore 1 November 2015.

Thus, one has to conclude that no firm recommendations are available. In addition, one should not forget that an increase in consumption financed through a decrease in savings should consider the social cost of reduced savings. Such element is usually neglected. Actually, it seems that there is little interest among economists for the social cost of reduced savings. This probably deserves investigations that go beyond the scope of this chapter.

Based on estimates of expenditures, one can compute the economic impact, in terms of increased added value:

$$VA = S \times \tau \times M \times \theta$$

With

S, additional spending
τ, share of initial spending directed to domestic providers
M, multipliers
θ, added value rate

We review, in turn, these various parameters.

The share of local expenditure, τ, is difficult to evaluate. Reasonably, it can be close to 100% for local population, but should be somehow lower for foreigners. For instance, a tourist who buys a package abroad leaves some expenditure outside of the hosting country in the first round; this can reach 39% of expenditures if one considers some evidence available for tourism in Italy (Confturismo CISET, 2014). This however deals only with tourists buying a package. The leakage is arguably much lower for other categories. In the absence of more convincing and comprehensive data, we use the simplifying assumption that $\tau = 1$, an assumption that may overestimate the impact of the event.

As far as the multiplier M is concerned, we use the value proposed in existing Expo studies. Again, this may be an overestimate. When prices increase during the event (this is likely to be the case for hotels) technical coefficients decrease compared with those under normal conditions.

As far as added value rates are concerned, they can be based on existing impact studies that indicate a value of 43% (Airoldi et al., 2010; Dell'Acqua, Morri and Quaini, 2013).

Table 9.3 Increased added value in Italy generated by visitors' expenditures (mln €)

Added	Italians	Foreigners		Total	Hypothetical effect
	Stayers	Additional	Extenders		Higher consumption propensity
Expenditure	130	452	330	912	562
Gross production	351	1,220	891	2,463	1,516
Added value	149	519	379	1,048	645

Source: Our elaboration.

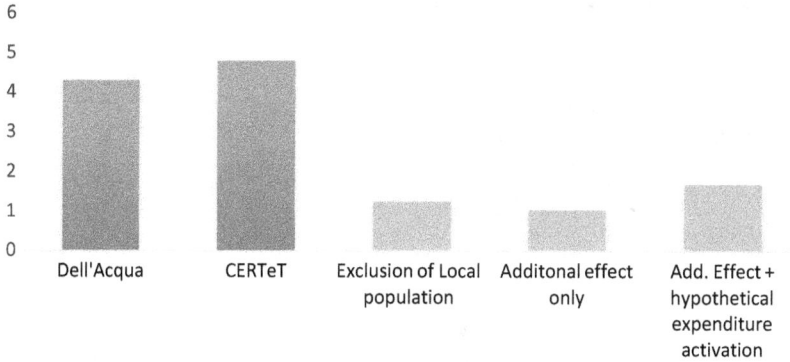

Figure 9.3 Comparison among estimation methods (added value in bln euro).

The resulting estimates are presented in Table 9.3. This table summarizes visitor expenditures, gross production and added value. The table distinguishes a first category of effects that are fairly certain for 1 billion euro of added value and other ones, based on activation of local consumption, that are more speculative for 0.6 billion euros. A comprehensive, yet speculative, estimate reaches 1.7 billion euro increase in added value. The various overestimation factors that we present in the two preceding paragraphs indicate that the real impact could be lower.

Figure 9.3 compares the outcome of the various existing estimations of the impact of Expo 2015. On the left, we have estimates by dell'Acqua *et al.* (2013) and Airoldi *et al.* (2010). On the right, various alternative estimates consider the effect of substitution. This relates to exclusion of the local population, and two of our estimates: a realistic one based on the "fairly certain effects" and a more speculative one "considering expenditure activation". Note that, in considering the more optimistic assumption, economists should also consider the shadow cost of reduced savings. This is recalled by the line labelled "shadow cost of reduced savings".

On the whole, it appears difficult to believe in estimates larger than two billion euros added value. Thus, the various estimates that correct for the impact of substituted expenditure converge to a sizeable reduction of the estimated impact.

The regional impact and inter-regional redistribution of wealth

The impact can also be measured at a regional scale. This is performed by Blake in a Computable General Equilibrium (CGE) (Blake, 2005), by McHugh in a Cost Benefit Analysis (CBA) (McHugh, 2006) and by Barget and Gouguet in an economic impact study (Barget and Gouguet, 2010). Others, like BOP,

Table 9.4 Additionality of visitors in BOP study

Visitors types	Edinburgh	Scotland
Local	3.0	1.1
Visitors from elsewhere in Scotland	85.1	1.6
Visitors from outside in Scotland	79.5	71.6

Source: BOP consulting (2011).
Original legend in quoted publication: "proportion of visitors to the Edinburgh festivals by place of origin, whose expenditure is additional to Edinburgh and Scotland, 2010".

proceeded with an impact study based on questionnaires. They find that only a tiny part of the visitors is additional for the territory of interest (Table 9.4). A question is then whether all territories (in this case Scotland versus Edinburgh) took advantage of the series of event, or whether the drain of wealth exerted by the main place of attraction has depleted the surrounding regions. Actually, events may have two contradictory effects: on the one side, a spillover tourist expenditure benefit for the surrounding regions, or a shift of Scottish consumer expenditures to Edinburgh on the other side. In the specific case, it seems the set of events was beneficial to both territories.

The analyst's conclusions run as follows:

Over 2010, the Edinburgh Festivals are estimated to have generated:

- new output of £245m in Edinburgh and £261m in Scotland
- £59m in new income in Edinburgh and £82m in Scotland
- supported 5,242 new FTE jobs in Edinburgh and 4,917 in Scotland.

(BOP, 2011)

How does it translate to our specific Expo case? An inhabitant of the Lazio region who says he would otherwise be at home the day of the visit is not additional for Italy but is so for Lombardy. A German tourist who says that without the Expo he would have stayed in another Italian region is also additional for Lombardy but reduced his stay in the rest of Italy. Taking this into account, 53% of the visits are additional for Lombardy. This figure is eight times larger than the corresponding figure for Italy. This already provides an order of magnitude on how the geographical distribution of the economic impact may be large compared with the impact at the national scale (see Table 9.5).

In addition, to increase the understanding of respondents' data we included some verification questions (with answers on a Likert scale from 0 to 10). For instance:

I came to the Expo today but, on the whole, I did not increase my stay in Italy (I would have come the same number of days).
Due to the Expo, I have increased the duration of my stay in Italy.

Table 9.5 Additionality of visitors for Lombardy

Origin		%	If I would not have come today to the Expo, I would have been…				% per origin		% of total visitors*	
			Lombardy	Rest of Italy	Home, work	Abroad	Substi-tutive	Addi-tional	Substi-tutive	Add
Rest of the world		15%	31%	16%	29%	23%	31%	69%	5%	10%
Lombardy		37%	25%	2%	72%	1%	97%	2%	36%	1%
Italy		48%	13%	32%	54%	1%	13%	87%	6%	42%
			Total general						47%	53%

Legend:

X%	% Additional
Y%	% Substitutive

To derive analytical results on the impacts on Lombardy versus the rest of Italy, we used the same conceptual framework given in the previous section. We compute:

- The impact of Lombard stayers
- The impact of additional Italians in Lombardy (who would have been in another region that day)
- The impact of Italians who extend their stay in Lombardy (they would have come to Lombardy in all cases but extended their stay)
- The impact of additional foreigners (they would not have been in Lombardy this day)
- The impact of foreigners who extend their stay in Lombardy (we assume the extra days spent in Italy have been spent in Lombardy; this could be investigated deeper in future research).

The indirect and induced regional impacts we present have been calculated using the regional multiplier provided in the bidding file (Comitato di candidatura, 2006); using other available multipliers (Dell'Acqua et al. 2013) would have produced an higher impact (by 68% to be precise).

BOX 9.2 COMPUTATION OF THE ADDITIONAL EXPENDITURE FOR LOMBARDY

The additional expenditure spent in Lombardy results from three sources:

1   Non-Lombard Italians:

They make additional trips to Lombardy,

$$N \times \sigma_{nl} \times \alpha_{nl} \times (g_{nl} \times S_{nl})$$

or prolong their stay:

$$N \times \sigma_{nl} \times (1 - \alpha_{nl}) \times (g'_{nl} \times S_{nl})$$

2   Lombards:

Some Lombards decide to stay in Lombardy rather than spending their money in other places

$$N \times \sigma_l \times \alpha_l \times E_l$$

3   Foreigners:

They make additional trips to Lombardy.

$$N \times \sigma_s \times \alpha_s \times g_s \times S_s$$

or prolong existing trips:

$$N \times \sigma_s \times (1 - \alpha_s) \times g'_s \times S_s$$

With,

N Total visitors
g_{nl} Overnight stays in Lombardy for non-Lombard Italians making an additional trip (nights/visit)
g'_{nl} Additional overnight stays in Lombardy for non-Lombard Italians that prolong their trip (night/visit)
g_s Average stay duration for each foreigner (night/visit)
g'_s Additional overnight stays for non-additional foreigners staying longer (nights/visit)
S_{nl} Daily expenditure of non-Lombard Italians (euro/day)
S_s Daily expenditure of foreigners (euro/day)
E_l Alternative expenditure for additional Lombards
σ_{nl} Share of non-Lombards Italians among the whole visitors
σ_s Share of foreigners (15%, our survey)
σ_l Share of Lombards among total visitors
α_l Share of additional among Lombard visitors (our survey)
α_{nl} Share of additional among Non-Lombard visitors (our survey)
α_s Share of additional among foreigner visitors (our survey)

To this, one may also want to add the hypothetical activation of consumption from the local population: local population may have increased its consumption propensity compared with a no event scenario.

Table 9.6 Increased added value generated by visitors' expenditures in Lombardy (mln €)

	Lombard	Non Lombard		Foreigners		Total	*Hypothetical Effect*
	Stayers	*Visitors*	*Extenders*	*Visitors*	*Extenders*		*Increased consumption propensity*
Expenditure	110	757	99	393	259	1619	193
Gross production	215	1477	193	766	506	3157	376
Added value	91	629	82	326	215	1343	160

Source: Author

Our results (Table 9.6) suggest that the impacts of the event visitors are highly concentrated in the Lombardy region. This is no surprise but now comes with a quantification: the impact on Lombardy is roughly 1/3 higher than the total impact on Italy. This comes with the uneasy implication that added value decreased in the rest of Italy.

In interpreting this result we should not forget two caveats:

First, our computation of the impact on the rest of Italy is net of spillover effects: the extra added value in Lombardy also impacts other regions. This is present in the national matrix, but we don't have multiregional matrices, so that we cannot isolate this effect at the subnational level. One could then wonder whether these spillover effects could be relevant enough to have a positive impact on the rest of Italy (without Lombardy).

Second, we consider the impact of visitors' expenditures alone and do not consider for instance public infrastructure investment which should also impact territorial redistribution. This deserves a study of its own.

Conclusions

This research represents, to our best knowledge, the first application in Italy of methods aiming at distinguishing the substitutive part of the impact of visitors. It also considers the important issue of how an event shifts expenditure (at least visitors ones) from the whole territory to a single region.

At the national level, our research shows that proper consideration of substitution effects is crucial when assessing visitors' impacts. Such a correction could easily reduce the estimated impacts by more than 50%.

At the interregional level, it shows that proper consideration of shift in the expenditures of domestic population can have a dramatic effect. Figure 9.4 illustrates this mechanism. Based on our findings, close to 300 million euros of added value were shifted from other regions to Italy.

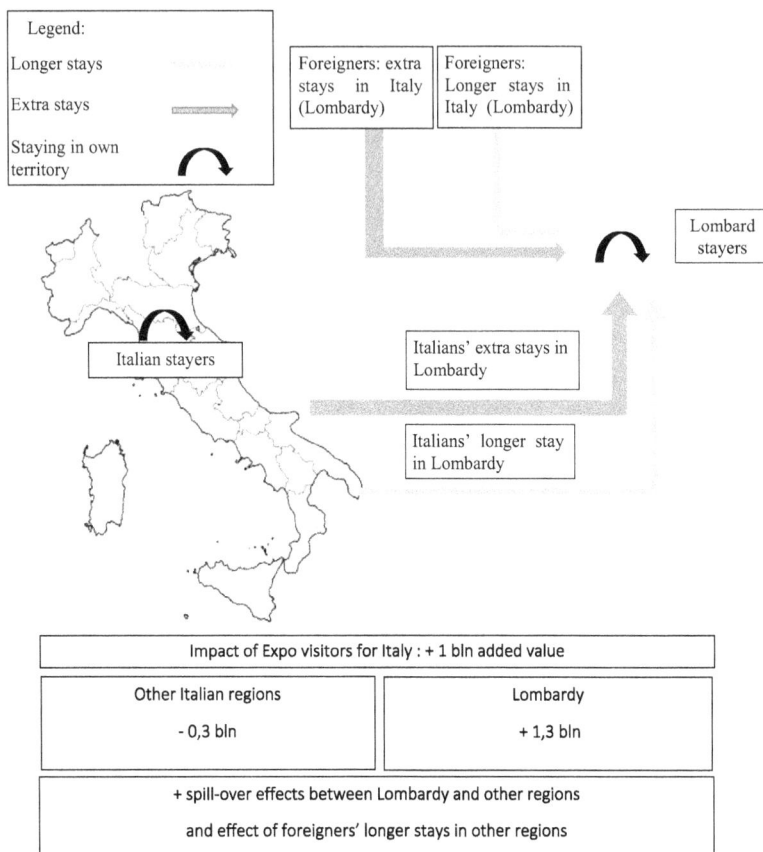

Figure 9.4 Value added redistributed by Expo 2015 through changes in expenditures from Italians and foreigners.

This calculation entails some simplifications. On the one side, it assumes that all increase in stay duration for non-additional foreigners is spent in Lombardy: an American that spends five days in Italy and says he would otherwise be somewhere else in Italy the day of its visits, and says he extended his stay in Italy for two days, is accounted for two extra days in Lombardy. This is reasonable but cannot be verified with certainty based on our survey. We can however note that the number of extra days in Italy for non-additional foreigners is quite low: 1.38, which means that once the extra days spent in Lombardy are deducted, the possibility that some extra days are available for other regions is limited. Even if extra days spent in Lombardy were kept to a minimum of one day, the impact on Lombardy added value would be reduced by a tiny fraction of 130 milion €, to the benefit of other regions. This shows that the proposed results are mostly robust to alternative assumptions more favorable to other regions.

Another limitation is that we do not consider economic spillover in the regional computation, but only in the national one. Part of the gross production necessary to satisfy the increase of final consumption in Lombardy will come from abroad and other regions. However the reasonable quantification of these spillover effects would likely not change the sign of the impact for the rest of Italy.

It may be worthwhile to derive as well some methodological proposals for future works.

About the survey instrument, we observed that such events provide excellent conditions for surveys. People queuing in a festivity atmosphere proved highly available to participate in the survey, with responses rates approaching 100% in most survey sessions.

This said we see room for improvement about evaluating the impact of the event by collecting more information on visitors' travel patterns. Actually, more information would be welcome on the number of days that each (group of) interviewed spent in various territories due to Expo. If one decided to go one week in Italy due to Expo and spent five days in Tuscany and two days in Lombardy this could be measured with more precisions and provide information for a more accurate estimate.

Our results suggest that the variable "number of days spent" in a given territory is crucial in determining the results. Average figures are quite low, slightly larger than 1, rarely exceeding two. But the final outcome is highly sensitive to this number, probably as much as to the value of the multiplier, which is frequently investigated with much detail.

Also, the more theoretical question of activation of expenditures (local population increasing its consumption propensity) deserves further investigation. On the one side empirical quantification would be useful: are observations consistent with this, often implicit, assumption? And its theoretical treatment through the marginal social costs of decreased savings deserves attention.

There is still much to do in order to better understand the real impacts of mega-events. This may come as a surprise for practitioners that believe a mechanical application of the model would be fit for purpose. "Net of substitution effects" is probably the keyword on this topic that requires further scientific edification. We hope to have laid a brick in this long-term construction process.

Acknowledgements

The present work has been achieved thanks to the input of Giorgia Pizziali when she was doing her master's thesis. Unfortunately, it has not been possible for Giorgia to take part in the writing of this chapter. We however are thankful to her and to other students that provided help in this work.

Appendix

Data on daily expenditures

The daily expenditure assumptions used in this study are based on CERTeT and Dell'acqua. These, in turn, are based on market studies by Eurisko and Bain & Company. The table presents the data that were used, as they are the most detailed among existing studies (Table 9A.1).

Table 9A.1 Daily expenditure assumption

Daily expenditure	Expected visitors (min)	Hotel nights per visitor	Hotel room night cost	Visitors / room	Lunch	Diner	Others
North Italy	6.1	0	110	1.4	15	25	26.25
Center	5.1	1	110	1.4	15	25	26.25
South Italy Professional/ Business	1.7	1	150	1.2	20	35	35.88
Abroad	7.1	4.3	120	1.4	20	35	30.63

Source: Trocciola (2014) based on CERTeT data.

References

Airoldi, A. *et al.* (2010) *L'impatto di EXPO 2015 nell'economia italiana.* CERTeT. Milano: Università Bocconi.

Baade, R. A., Baumann, R. and Matheson, V. A. (2008) 'Selling the game: Estimating the economic impact of professional sports through taxable sales', *Southern Economic Journal*, 74(3), pp. 794–810.

Barget, E. and Gouguet, J.-J. (2007) 'The total economic value of sporting events theory and practice', *Journal of Sports Economics*, 8(2), pp. 165–182. doi: 10.1177/1527002505279349.

Barget, E. and Gouguet, J.-J. (2010) 'La mesure de l'impact économique des grands événements sportifs. L'exemple de la Coupe du Monde de Rugby 2007, Measuring the economic impact of hallmark sporting events : The case of the Rugby World Cup 2007', *Revue d'Économie Régionale & Urbaine*, juin(3), pp. 379–408. doi: 10.3917/reru.103.0379.

Barget, E. and Gouguet, J.-J. (2011) 'De l'importance des dépenses des spectateurs étrangers dans l'impact touristique des grands événements sportifs', *Téoros. Revue de recherche en tourisme*, 30(30–2), pp. 105–119.

Bethlehem, J., Schouten, B. and Cobben, F. (2009) 'Indicators for the representativeness of survey response', *Survey Methodology*, 35.

Blake, A. (2005) 'Economic impact of the London 2012 Olympics'. Available at: https://opus.lib.uts.edu.au/handle/2100/994 (Accessed: 31 March 2017).

BOP (2011) *Edinburgh Festivals Impact Study Final Report*.

CERTeT. (2010). Expo Milano 2015 l'impatto sull'economia italiana (p. 5). *Milano* (Unpublished report).

Comitato di candidatura. (2006). Dossier di Candidatura Expo (p. 615). *Milano* (Unpublished report).

Confturismo CISET. (2014). Il turismo organizzato incoming. Dalla spesa del turista all'analisi della filiera, all'individuazione delle aree critiche (p. 80) (Unpublished report).

Crompton, J. L., Lee, S. and Shuster, T. J. (2001) 'A guide for undertaking economic impact studies: The springfest example', *Journal of Travel Research*, 40(1), pp. 79–87. doi: 10.1177/004728750104000110.

Dell'Acqua, A., Morri, G. and Quaini, E. (2013) *L'indotto di Expo 2015. Analisi di impatto economico*. Milano: SDA Bocconi.

Expo 2015 SpA (2016) 'Bilancio 2015. Nota integrativa alla situazione dei conti preliquidatoria con integrazione dei fatti successive'.

Felsenstein, D. and Fleischer, A. (2003) 'Local festivals and tourism promotion: The role of public assistance and visitor expenditure', *Journal of Travel Research*, 41(4), pp. 385–392. doi: 10.1177/0047287503041004007.

Getz, D. (1994) 'Event tourism: Evaluating the impacts', in Ritchie, J. R. B. et al. (eds) *Hospitality Research: A Handbook for Managers and Researchers*. New York: John Wiley & Sons, pp. 437–450.

Kruskal, W. and Mosteller, F. (1979) 'Representative sampling, III: The current statistical literature', *International Statistical Review / Revue Internationale de Statistique*, 47(3), pp. 245–265. doi: 10.2307/1402647.

Kruskal, W. and Mosteller, F. (1980) 'Representative sampling, IV: The history of the concept in Statistics, 1895–1939', *International Statistical Review / Revue Internationale de Statistique*, 48(2), pp. 169–195. doi: 10.2307/1403151.

McHone, W. W. and Rungeling, B. (2000) 'Practical issues in measuring the impact of a cultural tourist event in a major tourist destination', *Journal of Travel Research*, 38(3), pp. 300–303. doi: 10.1177/004728750003800313.

McHugh, D. (2006) *A Cost-Benefit Analysis of an Olympic Games*. Kingston, Ontario, Canada: Queen's Economics Department.

Nicolas, Y. (2007) *Les premiers principes de l'analyse d'impact économique local d'une activité culturelle*, p. 8.

Oxford Economics (2012) *The Economic Impact of the London 2012 Olympic & Paralympic Games. Oxford Economics*. Commissioned by Lloyds Banking Group, p. 52.

Preuss, H. (2005) 'The economic impact of visitors at major multi-sport events', *European Sport Management Quarterly*, 5(3), pp. 281–301. doi: 10.1080/1618 4740500190710.

Ryan, C. (1998) 'Economic impacts of small events: Estimates and determinants—a New Zealand example', *Tourism Economics*, 4(4), pp. 339–352.

Solberg, H. A., & Preuss, H. (2007). *Major Sport Events and Long-Term Tourism Impacts. Journal of Sport Management*, 21(2), 213–234. Human Kinetics, Inc. doi:10.1123/jsm.21.2.213

Taks, M. *et al.* (2011) 'Economic impact analysis versus cost benefit analysis: The case of a medium-sized sport event', *International Journal of Sport Finance*, 6(3), pp. 187–203.

Tyrrell, T. J. and Johnston, R. J. (2001) 'A framework for assessing direct economic impacts of tourist events: Distinguishing origins, destinations, and causes of expenditures', *Journal of Travel Research*, 40(1), pp. 94–100.

Vanhove, N. (2005) *The Economics of Tourism Destinations*. Oxford: Butterworth–Heinemann, ISBN 0-7506-6637-4.

Webber, D., White, S. and Smith, E. (2011) *Measuring Tourism Locally, Guidance Note Six: Event Analysis and Evaluation Edition No.: 1.0*, p. 21.

10 Destination Israel and Malawi wanderlusts

James Malitoni Chilembwe

Introduction

The term wanderlust encompasses different concepts (Mortimer, 2008); however, in this chapter, it refers to the desire to travel to another place either domestically or internationally to participate in an organised religious event or tourism event. The behaviour of visiting places of tourists' interest, mainly, of unknown experience, is not a new phenomenon. People have always travelled since the ancient days of world discovery and the trends continue without showing any signs of stoppage. Events such as group marathon, mount climbing, lake or water yachting, Christmas gatherings, New Year celebrations, cultural and music festivals, world cup and regional soccer events are organised yearly around the globe. Religious events are those events organised periodically or annually for people of a particular faith belief. However, this category of the event also includes tourism activities, hence incorporating the word event tourism. Religious events can also accommodate some elements of tourism events (Stausberg, 2012). Besides participation in actual festival events, people also take part in extra activities, for example, going on organised group tours. They usually travel in organised tourism events as per travellers' choice and motivations. According to Chimpweya (2014), in religious events, visiting places of religious functions accommodate many people, for example, the Mount Olive, the Chapel of the Ascension, Palm Sunday Road, Garden of Gethsemane. On contrast, tourism events are those events deemed secular. People travel to destinations merely for tourism consumption and leisure; however, the primary travel motivator is event participation. Examples of these are annual music festival events, sporting events, local or international cultural tourism events. These events attract many people, and they require a minimum number of people to participate when it is an organised event, marketed event and require more financial resources. Destination managers or event organisers also set objectives and goals to achieve during such events.

Critical issues in tourism event destination

Brida, Meleddu and Tokarchuk (2017) on their study 'economy on cultural events and Christmas markets' note the impact of the event through an increased economic value on tourists spending, shopping and spending time with local families. They suggest tourists' perception enhances their willingness to spend on local products. For example, a Christmas market is a cultural and retail event that creates a recreational value. Visitors attend to relax, to enjoy the Christmas atmosphere and to spend time. The consumer surplus of Christmas markets conveys benefits for the society as a whole. Recreation is higher for one-day visitors. Buning and Gibson's (2016) study describe the influence of travel conditions of preferred destination, event and travel characteristics in the context of an action sports event within the destination. They find the need for communities seeking to attract sport tourists as a form of sustainable tourism development to receive advice on how to organise events by incorporating entertainment and attractions. Activities available in the destination are much more critical to active event tourists, compared to exiting travel conditions (such as group, individual or with family members, long or short distance, travel styles). There are also preferences for different attractions in the host destination. The event image affects return to the destination in the future and hosting events from a destination management perspective. Regardless, religious events' needs demand tailored management to accommodate religious tourism experiences, event management and local tourism development (Bowdin, Allen, Harris, McDonnell & O'Toole, 2011; Raj, Walters & Rashid, 2013; Cerutti, 2015).

Typologies may fall under the umbrella of the events industry – three main categories are business events – festivals and fairs, cultural events, and sporting events. Besides, there are business events; cultural events; community events; outdoor events – entertainment, live music events, concerts and theatre shows; sports events, spectator sports; charity events; fundraising, voluntary sector; social life-cycle events; and special events (Bowdin et al., 2011). Events vary regarding their scale and complexity and the numbers of stakeholders involved. Major or mega-events have more objectives due to numerous related sub-events and stakeholders involved, compared to small local or community events which mainly link to particular geography (Bowdin et al., 2011). Typology of events according to Raj et al. (2013) are sports events, corporate events, musical events, cultural events, business events, political or government events, private events and religious events.

Collins, Jones and Munday (2009) state that the environmental impact of both day to day sports' activities and organisations' processes had received attention by agencies attempting to go for green sporting events. Events have a potential impact on local ecosystems, utilise reserves of irreplaceable natural capital and contribute to carbon emissions related to climate change. However, environmental externalities can be both positive and negative. Significant events can involve the significant physical redevelopment of the host destination – both for sports facilities and event-related transport improvements. There are also issues of climate change due to environmental degradation as a critical issue, requiring proper action by all stakeholders. Thus, it is essential to understand the influence of events in the creation of public policies from the tourism experience perspective. This involves the importance of connecting tourist desires with the destination experienceand the ability to organise around tourism policies. In fact, "the economy of tourism experience comes with the idea that the tourists cannot be passive consumers" (Dalonso, Lourenco, Remoaldo & Netto, 2014: 182). Consumers are co-creators, not just of their own experiences but also of the places visited through stories created or said and photographs shared with the audiences. Tourism events are not independent of critical issues in society. Therefore planners, event managers and all key actors should aim to provide participants with an unforgettable and unexpected experience. The context of the consumer in tourism presents characteristics that favour experience, and that leads us to reflect on the authenticity of products and services (Mkono, 2013). Any tourism activity can generate an experience, whether good or bad. However, the result depends on the main actors involved such as event managers, tourists' firms, local and regional organisations and local community as hosts. The consumer experience has a significant impact on the development of local tourism.

On tourism and event management education, the event management course has been part of tourism for such a long time as compared to hospitality programs. However, the arrival of event management programs further complicates the already complex and contested debate about the demarcation of the tourism and hospitality curriculum. The argument is that events and hospitality curriculum need to be very strong in practical orientation, equally tourism, though the degree of involvement varies widely (Dredge, Benckendorff, Day, Gross, Walo, Weeks & Whitelaw, 2013). The critical challenges confronting the successful holding event rely particularly on involving events personnel who are highly skilled, motivated and satisfied. Personnel with a high passion for their work can have a low perception of workloads and casualisation. Therefore, there is a need to have better policies that can support long-term workforce sustainability in events activities to minimise leakages of some competent graduates to another industry sector.

The importance of events is not only economical but also intangible benefits such as national pride, patriotism and country image. Fourie and Santana-Gallego (2011) note that it is critical to think of the cost of organising events regarding infrastructure, security, transportation, hosting because it may not always be offset by hosting events. The planned numbers may not turn up as projected, or sometimes the event may be overshadowed by other unforeseen factors such as sickness, death of event participants and low media publicity. For example, factors such as fear of influenza, natural and political disturbance can have a negative impact in numbers patronising the event (Sun, Rodriguez, Wu & Chuang, 2013). It is more critical thinking about the environmental sustainability of the hosting destination event.

Events provide several benefits to the society. One of the benefits of events is destination competitiveness. Other benefits not based on marketing orientation are community building to urban renewal, and cultural development to fostering national identities that can associate with or without tourism (Getz, 2008). Getz and Page (2016) add that events contribute to tourism development within the commercial area/sector. Events are motivators of tourism within the destination. Events add values to destination attractiveness and act as a critical marketing tool in the promotion of places to attract visitors spending. Events have become a core element of the destination system where accommodation, transport, attractions and ancillary services have been specifically developed to assist destinations provide tourism potential and capacity to enable adequate growth . Events can contribute to destination marketing to and to securing a competitive advantage in the marketplace and to economic benefits surrounding the region (Jogo, Dwyer, Lipman, van Lill & Vorster, 2010; Kelly & Fairley, 2018).

While hosting events is costly, a destination which is well marketed and patronised has the potential to leverage the cost (Kellett, Hed & Chalip, 2008). It can optimise limited resources (Kellette & Fairley, 2018). It can distribute events to a broader area (Fairley, Cardillo & Filo, 2016). It can create opportunities for entrepreneurs at local, regional and national levels and other concerned stakeholders in general (Beesley & Chalip, 2011), for example public and private associations or organisations within the sector. Leveraging strategies for a mega-event are not likely to affect the event itself. However, the use of the same strategies for small events may have some implications for the event. By extending to small events can affect the optimisation of tourism outcomes. Event leveraging is a strategic and proactive approach for formulating, maximising and distributing the potential benefits from an event (Kellett et al., 2008). The benefits from events can be economical or business (O'brien, 2006), social community (Kellett et al., 2008), sports participation (Misener, 2015; Weed et al., 2015) or tourism (O'Brien, 2006). Moreover, events are touted as a useful means of generating tourism in destinations (Getz, 2008; Boukas, Ziakas & Boustras, 2012). Therefore, destinations employ leveraging strategies to maximise the impacts of events tourism (Ziakas, 2013).

Management of the critical issue of festivals and events can create emotional displays, leading to emotions being fake and emotional toll exerted (Laing, 2018). The management may appear highly relevant in the festivals and events context. Given the high pressure of the environment, the preparedness of volunteer staff and the encounters between staff and attendees form a vital part of the quality of the festival or event going experience (Hyatt & Chard, 2013). There is a need to understand emotional labour to better deal with the delivery, staging of events and the experience of being a volunteer or attending an event. Examples of theoretical developments in the human resources management field that is useful for festival and event tourism are the concepts of burnout. Others are workplace stress, conservation of resources and workplace commitment.

Challenges linked to the sustainability of festivals, event tourism and destinations are also critical in event management. The sustainability of festivals and events is relatively well advanced. This includes the impacts of festivals and events on the sustainability of destinations, host communities, the planning of sustainable festivals and events. Other things to consider are the motivations of attendees at green events, attempts to use festivals and events to deliver sustainability messages and strategic objectives of the festival and event organisers that link to sustainability outcomes. Events may experience critical issues concerning the use of indigenous, traditional culture and rituals within festivals and events. Problems of authenticity, commodification and the potential exploitation that connect to more positive outcomes such as reconciliation and capacity building require further study and analysis in a variety of contexts. Liao (2016) in a study of space, memory and events adds that cultural productions and their essential events provide a space of experience that re-creates cultural meanings and reshapes memories. Tourism event stands at the nexus of tourism and event management and involves event planning, marketing in the pursuit of tourist and place marketing objective (Getz, 2008; Lucia, 2013). Investment decision-making in events becomes a core strategy for urban destinations that wish to improve their tourist development and their growth and requalification. Economic issues usually prevail in event strategy-making and portfolio building (Gertz, 2005). Events that gradually and regularly occur in the same location can help to raise/increase the identity of the place in which they are held, become a driver for image-making, and promote marketing and branding. Economic impact analysis is critical in events because it helps to develop event strategies and portfolios.

On traditional events, it is critical to have a governing body that sanctions for events and establishes and enforces standardised rules and regulations to follow during the production of the event (Mallen & Adams, 2013, 2017). The rules and regulations of the governing body specify elements such as the competition area, the number of participants, their dress and acceptable actions for participation. There are three critical characteristics of niche events. First, the event is created and adopted for a particular sport, recreation or tourism audience. Second, there does not have to be a traditional body that establishes time-honoured rules and regulations. However, the organising body may exist that may provide rules and regulations. Third, the event may exhibit recognisable traditional event components or may be conventional in its form.

The analysis of events portfolio is of direct relevance for event tourism planning and management as an increasing number of Destinations Management Organisation (DMOs) plan, manage and market a significant number of events (Getz, 2008; Ziakas, 2013). Getz (2005) describes the bundle of planned events in a destination in the form of the four-tier pyramid as local events, regional events, periodic 'hallmark' events and occasional mega-events. Events tourism can contribute to destination branding. Destination branding is a set of marketing activities to first support the creation of a name that readily identifies and differentiates a destination. Second, events can consistently communicate the expectations of memorable travel experiences. Third, events can serve to consolidate and rein, force the emotional connection between the visitor and destination and fourth reduce consumer search costs and perceive risk (Mariani and Giorgio, 2017). Collectively, these activities serve to create a destination image that positively influences consumer destination choice (Blain, Levy & Ritchie, 2005: 337, cited in Mariani & Giorgio, 2017). A destination image according to Bigne Alcaniz, Garcia and Sanz Blas (2009: 716) is "an idea, belief, feeling or attitude that tourists associate with the place". Destination branding process consists of three elements:

- The destination image identity (the image inspired in the tourism marketplace for the destination)
- Destination brand image (the actual brand image held by tourists)
- Brand positioning – consists of marketing communication activities to have brand identity overlap with brand image.

On historical events, McDowell and Skillen (2016) suggest that they should be looked at by stakeholders with three key categories: first owners and organisers, participants and spectators. The second is the relationship between the three key stakeholders and third the design of events through the use of ritual. On globalisation and critical events, the press and media should be in the forefront, putting the critical issues in events about the rewards and risks under a microscope, for example, financial crisis and climate change (Montessori, 2016). People do protests in major cities around the world as part of critical events to voice out their concerns and feelings about issues in society. Some events are emergency events, organised during sad times, for instance people always converge at various airports in times of plane crashes and loss of their relatives (Air France, March 2009, Malaysia Airways, July 2014, Lion Air, October 2018 and Ethiopian Airlines, 2019). People can also organise demonstration events to voice out their concerns in society. More recently, people converged at Hong Kong Airport during demonstrations, and similar demonstrations appeared in all cities in Malawi – effects of May 21 post-election 2019.

On the other hand, Raj and Bozonelos (2015) note on the 'pilgrimage experience and consumption of travel' that pilgrimage routes and experiences are increasingly becoming secularised, with their explicit promotion for tourism, leisure and cultural engagement. Despite that, the temporal world relationship between an individual and their beliefs can still act as a motivator, which influences an individual to visit religious sites. The visitation pattern depends on internal factors that are connecting the strengths of religious beliefs. Rajesh (2014) argues in 'trends of event tourism promotion in destinations' that event is anything that happens, as distinguished from anything that exists or an occurrence, especially one of great importance. The local community event is an activity established to involve the local population in shared experience of their mutual benefits. A special event is a unique moment in time with ceremony and ritual to satisfy event from two perspectives of customers and event managers. It is a one-time or frequently occurring event, outside regular programs or activities of the sponsoring or organising body. Such events are an opportunity for leisure, social, and cultural experience outside the normal range of choices or beyond everyday experience. Examples of special events are sporting events, corporate meetings, conventions, expositions, festivals, carnival and prize-giving ceremonies. The categories of events are according to the size of the events, for example, there are mega-events, regional events, major events and minor events. There is also classification according to geographical location, size of the population, the age of the population, organisations, the affluence of the community and organiser. The four broad categories of the events based on their purpose and objective are leisure events, cultural events, personal events and corporate events because they target and attract participants of different motives.

Several logistics requirements for event tourism destinations exist. For example, events need suitable venues, transportation facilities, hotels, event designers, staging, trained personnel in area of event management, operations, logistics, safety and security, crowd management modules, monitoring, control and evaluation system, health and medical support, emergency, technical expertise, event organisers, necessary facilities and amenities. Moreover, festivals provide communities with a way to celebrate their unique cultural traditions and attract tourist and local visitors in satisfactory and loyalty manner (Tanford & Jung, 2017). Festivals fall under the umbrella of event tourism, whereby people travel to destinations to attend specific events (Getz, 2008). Festivals span a variety of themes, including art, culture, food and beverage, music, religion and sports (Maeng, Jang & Li, 2016). Festivals provide economic, social and cultural benefits to the communities in which they occur (Grappi & Montanari, 2011), but there are also vital for destination marketing and to promote tourism (Chang, 2006). Regarding events, tourism has been adopted as a strategic approach throughout the destination management industry; events are recognised as an essential tourism product (Todd, Leask, & Ensor, 2017).

Event tourism is concerned with the production and marketing of events as motivators for tourism while also taking into consideration the value of events within destination management settings. Destination managers, along with event planners and producers, must be in a position to evaluate the tangible economic and marketing impacts of events, while attempting to gain an understanding of events' symbolic influences upon their hosts' destination image and brands (Getz, 2008; Getz & Page, 2016). More importantly, the managers are also increasingly adopting approaches to drive event tourism, hence creating unique selling points and differentiating their destinations from that of competitors (Getz, Svensson, Peterssen & Gunnervall, 2012). The literature considers hallmark events regarding their definition, function and form (Getz et al., 2012). Their purposeful development and management recognise them within tourism destinations, and their status can define longevity, quality of the brand image, significance and value. It is worthy to say that there are variations regarding categories, unlike mega or significant events; the scale of audience and media interest rarely recognises hallmark events. However, hallmark events are significant to the appeal and profitability of tourism destinations, and may be used as responses to seasonality.

Hallmark events can have an impact on the international recognition of their host destination, causing them to become tangible and symbolically embedded as permanent institutions within their community or culture (Ritchie, 1984 cited in Todd et al., 2017). It is so critical and challenging to manage events tourism in an environmentally sustainable way due to associated environmental issues and concerns when hosting events (Yuan, 2013). The environment is one of the central concerns for organising an event. Events do not only serve as tourist attractions, the catalyst for the development and renewal of destinations (Getz, 2008), but also cause significant environmental impacts (Collins, Flynn, Munday & Roberts, 2007). The damage of environmental assets that tourism ultimately depends upon will lead to severe negative economic impacts (Cater, 1995). Profit concerns drive policymakers and organisers in action. However, societal pressures are also increasingly interested in understanding how to incorporate environmental sustainability into the management of event tourism to reduce environmental impacts. Environmental sustainability of events tourism generally addresses the reduction and monitoring of environmental impacts of event organisation and delivery, the impacts of events on the community and the behaviours of event-goers (Getz, 2005; Smith-Christensen, 2009).

To minimise the impacts of planned tourism events in several ways, these complex dimensions of environmental sustainability should incorporate an event tourism management framework. First, the destinations should have systematic planning, development and marketing of events as tourist attractions, catalysts for other developments, image builders and animators of attractions and destination areas. Second, event tourism strategies should also cover the management of news and adverse events by organisers and sponsors. Third, event-goers, a market segment consisting of those people who travel to and attend events must be prioritised. Environmental resources provide essential ingredients for the construction of the infrastructure and production of the tourist product for the host community and event-goers. Event-goers act as tourist participants in events and gain experience that differs from a typical day of living. They pursue events for leisure, social and cultural purposes with high expectations on the quality of the environment (Yuan, 2013), which is key to provide tourist satisfaction, needs and wants; moreover the economic success of these destinations depends on it. Event tourism can be a national or regional development tool to allocate more re-sources in the budget, to construct infrastructure and to promote the desti-nation for attracting more tourists, considering environment as the essential cornerstone in tourism. Carter (1995) states that safeguarding the environ-ment is essential so that it any danger of being destroyed. While Rodella, Corbau, Simeoni and Utizi (2017) suggest the inclusion of carrying capacity issues as one way of crowd control to sustain destinations. Minimising the quality of the environment is essential in sustaining the attractiveness and capability of destination to host future events.

Event tourism and travellers to Israel

There are several religions in the world, but Christianity has a large fol-lowing (Chitsulo, 2017) with a large number of tourist arrivals in Israel. As such, there is a need for proper management of religious heritage sites and visitors to a destination like Israel (Olsen, 2006, 2009; Peaty, 2011). Such visits to places of cultural heritage and physical attraction are generally due to curiosity by believers and non believers alike. It is essential to seg-ment tourism consumers (Dolnicar & Kemp, 2009); to Israel as religious travellers if their travel links to religious events. Israel is renown as a key religious tourism destination (Timothy & Olsen, 2006; Gil & Curiel, 2008; Manfeld & McIntosh, 2009; Sharpley, 2009). However, Gilad (2017) laments that the number of all visitors to Israel is less than half as compared to other places of heritage and physical attractions, indicating that around 500,000–700,000 Christian pilgrims visit Israel per year. According to Gilad (2017), the total number of Christian tourists who dip into the Yardenit baptismal site in the Jordan River reach 350,000 every year.

However, Israel's religious tourist numbers appear far below when com-pared to other pilgrimages. Lourdes in France and Fatima Shrine in Portu-gal attract up to six million visitors a year while the Cathedral of Santiago de Compostela in Spain attracts one and a half million visitors a year, which is double than that of Israel (Gilad, 2017). Regardless of the ongoing con-flicts contributing to travel drawbacks, l, travel tours to Israel should attract significantly more visitors for its symbolic importance as the birthplace of Jesus Christ, where [He] physically walked, the places where Jesus performed miracles and addressed believers as well as his existing empty tomb, (Gilad, 2017). Chitsulo (2017) adds that Christians travellers who visit Israel return to their homes with both spiritual and tourism lessons. Residents have a passion catering to tourists needs; for example, they prepare water from a well which Jesus turned into wine. They sell clay used to mould pots and jars, and religious food and beverages; therefore, people are also motivated with religious tourism and entrepreneurial development (Rasul, Fatima & Sohail, 2016). There are a variety of hotels accommodation options availa-ble, with chains of restaurants. The provision of supporting services such as transport and guiding services helps visitors to feel at home and make repeat visits but also share the memories with colleagues who, in turn, get motivated too to visit the Holy Land of Israel.

According to Hall (2013: 370), travel can "sharpen one's sense of identity, including national identity, situated against the other". However, wanderlust religious travellers in case of Israel can encourage visitors to reflect upon their spiritual origin, values and faith. Hall (2013) adds that a fundamental aspect of the journey is the interactions that take place between the people of host destination such as the tour guides or religious leader tour guides as well as a reflection of spiritual life and the reality of what to observe, perform and get involved at the Holy sites. Moreover, there is an active link between personal memory and tourism, especially in the niche area of heritage tourism, where historical sites, cultural landscapes and collectively, are there to attract tourists (Marschall, 2012). Most of the decision choices regarding travel are driven by the sharing of personal memories of tourism, places of visit and experiences to various activities of host sites or events (Marschall, 2012). However, the reflection of different stories, values and beliefs in religious tourism and heritage is somehow very complicated because of variation in interpretation drives and market segments (Gelbman & Laven, 2016; Terzidou, Scarles & Saunders, 2018). Also, visitors to Israel have a lot to learn apart from gaining spiritual reflection. For example, the site attracts more people; yet the host has put in place some measures to manage visitor experiences. According to Tirca, Stanciulescu and Bacila (2010), managing visitor experience plays a key role in the continuous and long-lasting perception of the site and events that take place. In this case, the site or place including surrounding natural heritage should be well managed, thoroughly maintained and able to accommodate both religious and secular tourists (Brayley, 2010; Simone-Charteris & Body, 2010; Cerutti & Piva, 2015; Mustafa, 2015).

Wanderlusts and religious tourism events

Wanderlust is grounded on a broader concept of movement of landscapes such as pilgrimage, the physical anthropology of two-leggedness, mountaineering, walkathons, the formation of parks and walking meditation (Mortimer, 2008). The concept of wanderlust, or the impulse to travel, should not be ignored by travel marketers (Shields, 2011). Wanderlust is "the predisposition and impulsive urge to travel that permeates throughout all phases of the consumer's travel experiences" (Shields, 2011: 370). The models of tourism consumption indicate that before actual travel, there is always a decision-making process (Shields, 2011). Travel wanderlusts to Israel, who are typically religious tourists (Stausberg, 2012), visit several places. They pay a visit to the model of Jerusalem in Jesus days, Pool of Bethesda, the old city, Golgotha, Garden of Tomb, City of David, Pool of Siloam, Western Wall, Nazareth, Cana, Mount of Temptation, Jericho, Dead Sea and Capernaum (Face of Malawi, 2011; Chimpweya, 2014). They usually travel on an organised tour which mostly lasts for ten days to visit the places according to travellers' choice and motivations.

According to Chimpweya (2014), religious tourists also visit the birth-places of Jesus Christ, John the Baptist, Mount Olive, the Chapel of the Ascension, Palm Sunday Road, Garden of Gethsemane, the Church of All Nations, Mount Zion, the Church of Saint Peter in Gallicantu and the Upper Room. The pilgrims visit also takes the religious tourists to Mary's Well in Nazareth, which is the oldest well in Nazareth and also believed to be a well that Virgin Mary used to draw water to her house (Ron, 2009; Face of Malawi, 2011). They also visit the Church of Cana, where Jesus Christ had his first miracle by transforming water into wine at a marriage. They also visit the Wailing Wall, the Dome of the Rock, Al-Aqsa Mosque, St Anne's Church, Basilica of the Annunciation and walking on the street that Jesus walked while carrying his cross to Calvary (Face of Malawi, 2011). Gelbman and Laven (2016: 107) interpret heritage as "the meaning and values associated with everything we suppose has been down to us from the past". It also means cultural traditions and physical elements of the past that have been preserved and transmitted to the future generations of the society (Timothy & Body, 2006; Timothy, 2011; Francois, 2018).

Religious tourism is also associated with heritage foods (Weidenfeld & Ron, 2008; Nolan, 2010; Shinde, 2011). According to Ron and Timothy (2013), Christian tourists are curious about ancient Biblical food that appears more in the tourism landscape of both Jerusalem and some parts of Israel as an essential part of the pilgrimage experience associated with food. They further state that Christian pilgrims have developed an interest in local traditional food but also questioning what Jesus used to eat. Although the tourism industry seems to have limited responses to religious visitors, as Ron and Timothy (2013) observes, however, it is evident that the visitors want to know if local food they eat was the same that Jesus Christ ate. Perhaps, it is also one reason why Israel has its tourism links to religion.

The demand and growth in numbers encouraged private business firms to construct hotels and food chain restaurants to manage tourism sites by taking into consideration tourist needs. The Holy Land is open to both believers and non-believers. These believers and non-believers have several needs. While non-believers may require services in general, believers are strict about acquiring the experiences and reflection of their faith. Pilgrimages, festivals and religious events are some of the main motivating factors why people travel to sacred places (Blackwell, 2007; Cahener & Mansfeld, 2012; Shepherd, 2018). However, Terzidou et al. (2018) observe that there are complexities of religious tourism motivation. Although people travel together, however, their religious needs vary widely. Some may be interested in historical attractions related to their faith, while others may be interested in stories and heritage outside their religion. As such, a well-organised tour for the Christian tourists should include, among others, prayers conducted by religious groups as well as other leisure activities. In other words, there is a need to package other products that can make people have memories outside religion-focused, for example swimming, walking distances and climbing terraces.

According to Collins-Kreiner (2018: 8), "pilgrimage is journey resulting from the religious cause, externally and internally for spiritual purposes and internal understanding". Visitors to events in Israel have such kind of association in society due to the religious affiliation and cultural beliefs that have existed in the society for many years. However, of late, this pilgrimage mobility has experienced a shift, as it has become a pull factor not only for the religious travellers but also for secular visitors (Seaton, 2002; Stone, 2006). On the same vein, Collins-Kreiner (2015) notes that other than traditional destination pilgrimages, researchers are also getting interested in visiting secular historic shrines, spiritual festivals and sites, war memorials and graves including sporting activities at the destination. The places of pilgrimages perhaps have now turned into both religious and secular tourism events. However, there are also other motivations why people travel to such places as Israel, and thus it becomes difficult to categorise them as religious travellers or vacationers (tourists).

Moreover, there is a strong link between religious travellers and vacationers due to the revolution in religious tourism which can be applied either in sacred or a secular way when it comes to the social interaction at pilgrimage destination sites (Collins-Kreiner, 2018). There appears that modern pilgrimages for tourists, pilgrims, and local people are competing for resources such as infrastructure, car parking spaces around sacred places and transportation (Collins-Kreiner, 2018). Therefore, the chapter emphasises the focus and consideration on international visitation to pilgrimage destinations as wanderlust religious tourists or religious travellers. In this case, any form of travel away from home for either leisure or business for at least 24 hours or more is tourism; as noted earlier apart from requiring accommodation and transportation, these tourists also need local traditional food (Ron & Timothy, 2013). Pilgrimages change the economic, social and political landscape of tourism destinations around the world due to their scope and spatial influence (Karar, 2010; Henderson, 2011; Chilembwe, 2014; Liutikas, 2015). As Liutikas (2015) notes, pilgrimages are beneficial to local communities due to employment creation, as locals can sell various products to visitors that depict the history and attractions of the sites. Religious tourists buy these products to have something to show others at home and to have a long memory about the place linked to their religious faith. On a similar note, Henderson (2011) stated that approximately 11.6 million international religious arrivals travelled to Saudi Arabia in 2008, generating about 75% revenue from religious events. On the other hand, Mecca received 6.6 million religious tourists, suggesting that these events are powerful to contribute to the economic and social landscape if well managed (Henderson, 2011). Socially, pilgrimage destination events bring together people of either the same or different faith or beliefs together as people are interested in discussing a range of issues affecting the society, including their religious beliefs.

Niche tourism

Organising religious tours to places of religious interest is not a new phenomenon worldwide (Collins-Kreiner, 2018). The only difference is that around the world, religious tours are organised by specialised tour operators. In contrast, in Malawi, the previous religious tours were mostly organised by religious faith groups for various factors, for example, beliefs and values. As Gelbman and Laven (2016) observe, travellers always differ in their reflection of various stories, values and beliefs in religious tourism and heritage because of complex and variation in interpretation drives and market segments. Marschall (2012) adds that personal memory plays a significant role in tourism, especially in the niche area of heritage tourism, where historical sites and cultural landscapes are there to attract tourists. However, these strong linkages in values and beliefs to attract religious travellers to niche areas related to religious tourism are incorporated by modern private tour operators.

Moreover, globalisation and technological advancements continue to play a critical role in the creation of new ideas and developing new forms of tourism and niche tourism markets to maximise not only revenue for tour operators but also tourists' values and experiences (Stanis & Barbieri, 2013). Several examples of niche tourism that are not limited to whale watching, bird watching, elephant tracking, surf tourism, shark diving and volunteer tourism surfaced in the last decades. Raymond and Hall (2008) predict that niche market segments will continue to grow in the future. Tomazos and Butler (2010) note that nowadays many private tour operators have entered into the tourism industry due to the growth of niche segments. In Malawi, for example, tour operators are doing the work that was previously handled by regional organisations due to their inclination towards tours around religious beliefs and values. As Stanis and Barbieri (2013) echo, the growing interest and popularity of religious tourism bring in specialised tour operators to provide technical expertise and assistance to both individuals and interested groups willing to participate in religious tours. In the case of Malawi, the majority of travellers have links to travel within the African continent, particularly South Africa, Botswana and Zimbabwe due to closeness and cost benefits. Some travel as far as China, the United Kingdom, and the United States, mostly for business purposes though at times may also combine with leisure activities such as tour participation, visiting museums and other places of tourists' interests.

Interestingly, a new demand has recently emerged for outbound tourism to Israel-Jerusalem, specifically for religious travellers. This new niche market segment does not only attract people of faith groups like church leaders and their followers but also non-believers who are visiting the Holy Land.

Some do make visits for pure tourism landscapes and understand food tourism as observed by Shinde (2011). Besides the Holy Land, Malawian religious leaders and other believers were associated with visits to places of fellow religious common beliefs for many years. They mostly travelled and visited places such as Aberdeen, Mecca, Pittsburgh and Vatican City through partnership and pilgrimages. Tour operators are now focused on segments that were previously organised by religious faith groups, as it is now more convenient o organise tours professionally on behalf of interested groups (Chilembwe & Mweiwa, 2014, 2019). Faith groups and tour operators have established working partnerships with host tour operators to provide both effective escorted and hosted tours as well as marketing their tours (Chilembwe, Mweiwa & Mankhomwa, 2018). Earlier, there was only one organised tour a year. However, due to the increase in demand, the number of organised tours per year has kept on increasing. Equally, the number of tour operators interested in the niche market segment has also increased. In this context, the niche segment is a market emerging from already Malawian existing religious markets such as Vatican City, Pittsburgh, Aberdeen and Mecca tours. It is only recently that Israel tours appear more appealing to the Malawi wanderlusts.

Conclusion

There are several critical issues to take into consideration when organising events. Whether it is a religious or circular event, it cannot be successful if there are no essential requirements in place. Planning for transportation, accommodation, attractions, activities and supporting facilities are some of the fundamental things that can make organised tours and events successful. Events can be either small, mega, local regional, national or international (Getz, 2008). Despite what form of an event is, its success is dramatically dependent on the organisers of the event and also partly on the participants of the events. While organisers have a huge part to play in the events, however, without responsible participants who care about the sustainability of environments may be a setback to an organised event.

Notably, wanderlusts seeking new experiences of the unknown should pay attention to the rules of the events (Mortimer, 2008). In Malawi wanderlust to destination Israel case, the experience takes religious event travellers in the form of religious tourism, although the majority of Malawi's wanderlusts travel more at the regional level, but more recently extended to the destination – Israel. This destination, as a new market segment for Malawians, adds on the existing list of other religious destinations such as Vatican City, Pittsburgh and Aberdeen, also more familiar with religious travellers from Malawi. Therefore, travel to Israel includes both religious and non-religious Malawian travellers.

However, many religious travellers (tourists) dominate the visit for the fact that religion and faith motivate and connect the majority (Chimpweya,

2014). Religious beliefs take centre stage in the drive of travellers to strengthen their faith, personal experience, meditation and spiritual life in various groups. Events though look ordinary are not only connected to religious events but also to soft and adventure trips. They involve walking distances, swimming, sharing experiences, meditation, spiritual life reflection and imitation. Travel comprises of an adventure, experience and education, all enrolled in one. Travellers engage in religious tourism for the fact that they spend two to three weeks away from their homes. They pay for accommodation and local services but also buy some memorable or great products to take home is what makes the whole travel experience called 'religious tourism'.

The Holy Land in Israel is open to all regardless of religious affiliation or not. Acceptance of all interested people incorporates those that are only interested in food tourism. Besides adventure activities, religious tourists seek ancient Biblical food. Nolan (2010) and Ron and Timothy (2013) state that it is common to seek associated food in the tourism landscape all over the world as it appears part of the pilgrimage. Others call it 'heritage food' in Israel (Weidenfeld & Ron, 2008; Shinde, 2011). Thus, the demand and growth in the numbers of both religious tourists and secular tourists encourage the private sector to construct accommodation facilities. The availability of hotels and food chain restaurants as well as provision of local transport and tour guiding services is one way of managing tourism sites as part of the well-being of tourists. There should be also provision of related tourist welfare like physical fitness activities for sports, mountaineering and bicycle riding, jogging and swimming, and walking around the valleys during an adventure. The extension of visitors' facilities suggests trends of sacred places or Holy Land visits shifting from 'exclusively religious people' to 'inclusively non-religious people'.

Also, the interest of Malawi wanderlusts and people, in general, to travel to Israel (of a non-religious segment) is a perfect example of how Israel accommodates and hosts visitors. The establishment of a joint arrangement of organised tours with tour operators and religious organisations appears to contribute to Israel's destination marketing as well as to the increasing demand in the niche segment (Chilembwe & Mweiwa, 2014). Although regional organisations were the first to arrange religious tours, tour operators can organise tours professionally by collaborating with private tour operators in Israel. Through such partnerships, they make tailor-made tours, for both religious and secular travel groups, and provide escorted and hosted tour guiding and interpretation services.

References

Beesley, L.G., & Chalip, L. (2011) Seeking (and not seeking) to leverage mega-sport events in non-host destinations: The case of Shanghai and the Beijing Olympic. *Journal of Sport & Tourism*, 16(4), 323–344.

Bigne Alcaniz, E., Garcia, I.S., & Sanz Blas, S. (2009) The functional-psychological continuum in the cognitive image of a destination: A confirmatory analysis. *Tourism Management*, 30(5), 715–723.

Blackwell, R. (2007) Motivation for religious tourism: Pilgrimages, festivals and events. In: Raj, R., & Morpeth, N.D. (Eds.) *Religious Tourism and Pilgrimage Management: An International Perspective* (pp. 35–47). Wallingford: CAB International.

Blain, C., Levy, S.E., & Ritchie, J.R.B. (2005) Destination branding: Insights and practices from destination management organisation. *Journal of Travel Research*, 43, 328–338.

Boukas, N., Ziakas, V., & Boustras, G. (2012) Olympic legacy and cultural tourism: Exploring the facets of Athens' heritage. *International Journal of Heritage Studies*, 19(2), 203–228.

Bowdin, G., Allen, J., Harris, R., McDonnell, I., & O'Toole, W. (2011) *Events Management*, (2nd Ed). Oxford: Butterworth-Heinemann.

Brayley, R.E. (2010) Managing sacred sites for tourism: A case study of visitor facilities in Palmyra, New York. *Tourism*, 58(3), 289–300.

Brida, J.G., Meleddu, M., & Tokarchuk, O. (2017) Use value of cultural events: The case of the Christmas markets. *Tourism Management*, 59, 67–75.

Buning, R.J., & Gibson, H.J. (2016) The role of travel conditions in cycling tourism: Implications for destination and event management. *Journal of Sport & Tourism*, 20(3–4), 175–193.

Cahener, L., & Mansfeld, Y. (2012) A voyage from religiousness to secularity and back: A glimpse into 'Heredi' tourists. *Journal of Heritage Tourism*, 7(41), 301–321.

Carter, E. (1995) Environmental contradictions in sustainable tourism. *The Geographical Journal*, 161(1), 21–28.

Cerutti, S. (2015) Religious tourism and event management: An opportunity for local tourism development. *International Journal of Religious Tourism and Pilgrimage*, 3(1/8), 54–65.

Cerutti, S., and Piva, E. (2015) Religious tourism and event management: An opportunity for local tourism development. *International Journal of Religious Tourism and Pilgrimage*, 3(1), 54–65.

Chang, J. (2006) Segmenting tourists to aboriginal cultural festivals: An example in the Rukai tribal area, Taiwan. *Tourism Management*, 27(6), 1224–1234.

Chilembwe, J.M. (2014) Examination of social cultural impacts of tourism in Chembe Village in Mangochi District, Malawi. *International Journal of Business Quantitative Economics and Applied Management Research*, 1(1), 61–92.

Chilembwe, J.M., & Mweiwa, V.R. (2014) Tour guides: Are they developers and promoters - Case study of Malawi. *Impact Journals: International Journal of Research and Business Management*, 2(9), 29–46.

Chilembwe, J.M., & Mweiwa, V.R. (2019) Responsible travel and tourism adventure: Evidence from Malawi as a tourist destination. In: Sharma, A. (Ed.), *Sustainable Tourism Development: Futuristic Approaches* (pp. 31–53). Florida: Apple Academic Press.

Chilembwe, J.M., Mweiwa, V.R., & Mankhomwa, E. (2018) Role of tour operators in destination tourism marketing in Malawi. In: Camilleri, M.A. (Eds.), *Strategic Perspectives in Destination Marketing* (pp. 295–321). Hershey, PA: IGI Global Publishers.

Chimpweya, J. (2014, 26 October) *Synod Visits Israel.* Retrieved on 20th March 2018 from, https://mwnation.com/synod-visits-israel/.

Chitsulo, M. (2017, 04 November) *Christian Pilgrims Return with Spiritual, Tourism Lessons.* Retrieved on 13th March 2018 from, https://times.mw/christian-pilgrims -return-with-spiritual-tourism-lessons/.

Collins, A., Flynn, A., Munday, M., & Roberts, A. (2007) Assessing the environmental consequences of major sporting events: The 2003/04 FA Cup Final. *Urban studies*, 44(3), 467–476.

Collins, A., Jones, C., & Munday, M. (2009) Assessing the environmental impacts of mega sporting events: Two options? *Tourism Management*, 30, 828–837.

Collins-Kreiner, N. (2015) Dark tourism as/is pilgrimage. *Current Issues in Tourism*, 19(12), 1185–1189.

Collins-Kreiner, N. (2018) Pilgrimage-tourism: Common themes in different religions. *International Journal of Religious Tourism and Pilgrimage*, 6(1), 8–17.

Dalonso, Y., Lourenco, J.M., Remoaldo, P.C., & Netto, A.P. (2014) Tourism experience, events and public policies. *Annals of Tourism Research*, 64, 163–184.

Dolnicar, S., & Kemp, B. (2009) Tourism segmentation by consumer-based variables, In: Kozak, M., and Decrop, A. (Eds.), *Handbook of Tourist Behaviour Theory and Practice* (pp. 177–194). New York: Routledge.

Dredge, D., Benckendorff, P., Day, M., Gross, M.J., Walo, M., Weeks, P., & Whitelaw, P.A. (2013) Drivers of change in tourism, hospitality, and event management education: An Australian perspective. *Journal of Hospitality and Tourism Education*, 25(20), 89–102.

Face of Malawi (2011, 30 May) *Malawi Catholics on Pilgrimage to Uganda, Israel.* Retrieved on 20th January 2018 from, https://faceofmalawi.com/2011/05/malawi -catholics-on-pilgrimage-to-uganda-israel/.

Fairley, S., Cardillo, M., & Filo, K. (2016) Engaging volunteers from regional communities: Non-hot city resident perceptions towards a mega-event and the opportunity to volunteer. *Event Management*, 20(3), 443–447.

Fourie, J., & Santana-Gallego, M. (2011) The impact of mega-sport events on tourist arrivals. *Tourism Management*, 32, 1364–1370.

Francois, T. (2018) Western travellers in the Land of the Buddha legitimising travel through the religionification of Tourism. *International Journal of Religious Tourism and Pilgrimage*, 6(1), 33–46.

Gelbman, A., & Laven, D. (2016) Re-envisioning community-based heritage tourism in the old city of Nazareth. *Journal of Heritage Tourism*, 11(2), 105–125.

Getz, D. (2005) *Event Management and Event Tourism,* New York: Cognizant.

Getz, D. (2008) Event tourism: Definition, evolution and research. *Tourism Management*, 29(3), 403–428.

Getz, D., & Page, S. (2016) Progress and prospects for event tourism research. *Tourism Management*, 52, 593–631.

Getz, D., Svensson, B., Peterssen, R., & Gunnervall, A. (2012) Hallmark events: Definitions, goals and planning process. *International Journal of Event Management Research*, 7(1/2, 47–67.

Gil, A.R., & Curiel, J. (2008) Religious events as special interest tourism, a Spanish experience. *PASOS Revista De Turismo Y Patrimonio Cultural*, 6(3), 419–433.

Gilad, M. (2017, 16 May) *A Holy Mastery: Why Aren't Christian Pilgrims Visiting the Homeland of Jesus? Experts Put the Potential of Christian Faith Tourists at 10 Million a Year, Compared with 700,000 at Present. What's Israel Is Doing*

Wrong? Retrieved on 7th March 2018 from, https://haaretz.com/israel-news/. premium.MAGAZINE-the-mystery-of-israels-missing-christian-pilgrims-1.5472618

Grappi, S., & Montanari, F. (2011) The role of social identification and hedonism in affecting tourist re-patronising behaviours: The case of Indian festival. *Tourism Management*, 32(5), 1128–1140.

Hall, L. (2013) My Wanderlust is not yet appeased: Ellis Bankin and motorcycle touring in 1930s Australia. *Studies in Travel Writing*, 17(4), 368–383.

Henderson, J.C. (2011) Religious tourism and its management: The Hajj in South Arabia. *International Journal of Tourism Research*, 13, 541–552.

Hyatt, C., & Chard, C. (2013) Facilitating quality in event management, In: Mallen, C., & Adams, L.J. (Eds.), *Event Management in Sports, Recreation and Tourism: Theoretical and Practical Dimensions* (pp.181–197). London: Taylor & Francis Group.

Jogo, L., Dwyer, L., Lipman, G., van Lill, D., & Vorster, S. (2010) Optimising potential of mega-events: An overview. *International Journal of Event and Festival Management*, 1(3), 220–237.

Karar, A. (2010) Impact of pilgrim tourism at Haridwar. *Anthropologist*, 12(2), 99–105.

Kellette, P., Hede, A.M., & Chalip, L. (2008) Social policy for sport events: Leveraging (relationships with) teams from other nations for community benefit. *European Sport Management Quarterly*, 8(2), 101–121.

Kelly, D.M., & Fairley, S. (2018) What about the event? How do tourism leveraging strategies affect small-scale events? *Tourism Management*, 64, 335–345.

Laing, J. (2018) Festival and event tourism research: Current and future perspectives. *Tourism Management*, 25, 165–168.

Liao, D.Y. (2016) Space and memory in the Huashan event. In: Lemond, I.R., & Platt, L. (Eds.), *Critical Event Studies: Approaches to Research* (pp. 109–130). London: Palgrave Macmillan.

Liutikas, D. (2015) Religious landscape and ecological ethics: Pilgrimage to the Lithuanian Calvaries. *International Journal of Religious Tourism and Pilgrimage*, 3(1), 12–24.

Lucia, M.D. (2013) Economic performance measurement systems for event planning and investment decision making. *Tourism Management*, 34, 91–100.

Maeng, H.Y., Jang & Li, J.M. (2016) A critical review of the motivational factors for festival attendance based on meta-analysis. *Tourism Management Perspectives*, 17, 16–25.

Mallen, C., & Adams, L.J. (2013) Traditional and niche events in sports, recreation and tourism. In: Mallen, C., & Adams, L.J. (Eds.), *Event Management in Sports, Recreation and Tourism: Theoretical and Practical Dimensions* (pp.1–9). London: Taylor & Francis Group.

Mallen, C., & Adams, L.J. (2017) *Event Management in Sports, Recreation and Tourism: Theoretical and Practical Dimensions* (3rd Ed.). Abingdon, OX: Routledge.

Manfeld, Y., & McIntosh, A. (2009) Spiritual hosting and tourism: The host-guest dimension. *Tourism Recreation Research*, 34(2), 157–168.

Mariani, M.M., & Giorgio, L. (2017) The "Pink Night" festival revisited: Mega-events and the role of destination partnerships in staging event tourism. *Annals of Tourism Research*, 62, 89–109.

Marschall, S. (2012) 'Personal memory tourism' and a wider exploration of the tourism-memory nexus. *Journal of Tourism and Cultural Change*, 10(4), 321–335.

McDowell, M.L., & Skillen, F. (2016) The rewards and risks of historical events studies research. In: Lemond, I.R., & Platt, L. (Eds.), *Critical Event Studies: Approaches to Research* (pp. 87–107). London: Palgrave Macmillan.

Misener, L. (2015) Leveraging parasport events for community participation: Development of a theoretical framework. *European Sport Management Quarterly*, 15(1), 1–22.

Mkono, M. (2013) African and western tourists: Object authenticity quest? *Annals of Tourism Research*, 41, 195–214.

Montessori, N.M. (2016) CDA, Critical events and critical events studies: How to make sense of critical events in a society of radical change. In: Lemond, I.R., & Platt, L. (Eds.), *Critical Event Studies: Approaches to Research* (pp. 131–147). London: Palgrave Macmillan.

Mortimer, N. (2008) Wanderlust: A history of walking, time and mind. *The Journal of Archaeology, Consciousness and Culture*, 1(3), 389–390.

Mustafa, M.H. (2015) Tourism development at the Baptism site of Jesus Christ, Jordan: residents' perspective. *Journal of Heritage Tourism*, 9(1), 75–83.

Nolan, D. (2010) *A Food Lover's Pilgrimage to Santiago de Compostela: Food, Wine and Walking along the Camino through Southern France and the North of Spain.* Melbourne: Penguin.

O'brien, D. (2006) Event business leveraging the Sydney 2000 Olympic games. *Annals of Tourism Research*, 33(1), 240–261.

Olsen, D. H. (2006) Management issues for religious heritage attractions, In: Timothy, D.J., and Olsen, D.H. (Eds.), *Tourism Religion and Spirituality Journeys* (pp. 104–119). London: Routledge.

Olsen, D.H. (2009) 'The strangers within our gates': Managing visitors at Temple Square. *Journal of Management, Spirituality and Religion*, 6(2), 121–139.

Peaty, D. (2011) Sacred sites, conservation and tourism. *Ritsumeikan Seisakukagaku*, 18(3), 193–199.

Raj, R., & Bozonelos, D. (2015) Pilgrimage experience and consumption of travel to the city of Makkah for Hajj ritual. *International Journal of Religious tourism & Pilgrimage*, 3(1/6), 37–45.

Raj, R., Walters, P., & Rashid, T. (2013) *Event Management: Principles and Practices.* London: Sage Publications.

Rajesh, R. (2014) Issues and trends of event tourism promotion in destinations: Puducherry, an empirical study. *International Journal of Safety and Security in Tourism and Hospitality*, 1(6), 26–41.

Rasul, F., Fatima, U., & Sohail, S. (2016) Religion tourism and entrepreneurial development: A case study Hazrat Data Ganj Bakhsh Shrine. *South Asian Studies*, 31(1), 275–289.

Raymond, E.M., and Hall, C.M. (2008) The development of cross-cultural (mis) understanding through volunteer tourism. *Journal of Sustainable Tourism*, 16(5), 530–543.

Ritchie, J.R.B. (1984) Assessing the impact of hallmark events: Conceptual and research issues. *Journal of Travel Research*, 23(1), 2–11.

Rodella, I., Corbau, C., Simeoni, U., & Utizi, K. (2017) Assessment of the relationship between geomorphological evolution, carrying capacity and users' perception: Case studies in Emilia-Romagna (Italy). *Tourism Management*, 59, 7–22.

Ron, A.S. (2009) Towards a typological model of contemporary Christian travel. *Journal of Heritage Tourism*, 4(4), 287–297.

Ron, A.S., & Timothy, D.J. (2013) The land of milk and honey: Biblical foods, heritage and Holy Land tourism. *Journal of Heritage Tourism*, 8(2–3), 234–247.

Seaton, A.V. (2002) Thanatourism's final frontiers? Visits to cemeteries, churchyards and funerary sites as sacred and secular pilgrimage. *Tourism Recreation Research*, 27(2), 27–33.

Sharpley, R. (2009) Tourism, religion and spirituality, In: Jamal, T., and Robinson, M. (Eds.), *The Sage Handbook of Tourism Studies* (pp. 237–253). Thousand Oaks, CA: Sage Publications.

Shepherd, R.J. (2018) When sacred space becomes a heritage place: Pilgrimage, worship, and tourism in contemporary China. *International Journal of Religious Tourism and Pilgrimage*, 6(1), 33–46.

Shields, P.O. (2011) A case for wanderlust: Travel behaviour of college students. *Journal of Travel and Tourism Marketing*, 28(4), 369–387.

Shinde, K.A. (2011) Placing communitas: Spatiality and ritual performances in Indian religious tourism. *Tourism*, 59(3), 335–352.

Simone-Charteris, M.T., & Body, S.W. (2010) The development of religious heritage tourism in Northern Ireland: Opportunities, benefits, and obstacles. *Tourism*, 58(3), 229–257.

Smith-Christensen, C. (2009) Sustainability as a concern within events, in Raj, R. and Musgrave, J. (Eds.), *Event Management and Sustainability* (pp. 22–30). Cambridge, MA: CAB international.

Stanis, S.A.W., & Barbieri, C. (2013) Niche tourism attributes scale: A case of storm chasing. *Current Issues in Tourism*, 16(5), 495–500.

Stausberg, M. (2012) Spiritual tourism: Travel and religious practice in western society. By Alex Norman. *Numen*, 59(5–6), 618–620.

Stone, P.R. (2006) A dark tourism spectrum, towards a typology of death and macabre related tourist sites, attractions and exhibitions. *Tourism, An Interdisciplinary International Journal*, 52(2), 145–160.

Sun, Y., Rodriguez, A., Wu, J., & Chuang, S. (2013) Case study - Why hotel rooms were not full during a hallmark sporting event: The 2009 World Games experience. *Tourism Management*, 36, 469–479.

Tanford, S., & Jung, S. (2017) Festival attributes and perceptions: A mega-analysis of relationships with satisfaction and loyalty. *Tourism Management*, 61, 209–220.

Terzidou, M., Scarles, C., & Saunders, M.N.K (2018) The complexities of religious tourism motivations: Sacred places, vows and visions. *Annals of Tourism Research*, 70, 54–65.

Timothy, D.J. (2011) *Cultural Heritage and Tourism: An Introduction*. Bristol: Channel View Publications.

Timothy, D.J., & Body, S.W. (2006) Heritage tourism in the 21st century: Valued traditions and new perspective. *Journal of Heritage Tourism*, 1(1), 1–16.

Timothy, D.J., and Olsen, D.H. (Eds.). (2006) *Tourism, Religion and Spiritual Journeys*. London: Routledge.

Tirca, A., Stanciulescu, Chis, A., & Bacila, M. (2010) Managing the visitor experience on Romanian religious sites: Monasteries abbots' perceptions. *Management and Marketing*, 8(1), 5–16.

Todd, L., Leask, A., & Ensor, J. (2017) Understanding primary stakeholders' multiple role in hallmark event tourism management. *Tourism Management*, 59, 494–509.

Tomazos, K., & Butler, R. (2010) The volunteer tourist as 'hero'. *Current Issues in Tourism*, 13(4), 363–380.

Weed, M., Coren, E., Fiore, J., Wellard, I., Chatziefstathiou, D., Mansfield L., et al. (2015) Olympic game games and rising sport participation: A systematic review of evidence and interrogation of policy for a demonstration effect. *European Sport Management Quarterly*, 15(2), 195–226.

Weidenfield, A., & Ron, A. (2008) Religious needs in the tourism industry. *Anatolia: An International Journal of Tourism and Hospitality Research*, 19(2), 357–361.

Yuan, Y.Y. (2013) Adding environmental sustainability to the management of event tourism. *International Journal of Culture, Tourism and Hospitality Research*, 7(2), 175–183.

Ziakas, V. (2013) *Event Portfolio Planning and Management: A Holistic Approach.* Abingdon: Routledge.

11 City rebranding, social discontent and bidding for cultural events

Daniel Barrera-Fernández, Miguel Anxo Rodríguez-González and Marco Hernández-Escampa

Introduction

In recent years, competition among cities to host events with great international impact has increased dramatically. Major events not only constitute as a tool to attract visitors, but they also serve as a marketing strategy to offer a new image of the city to tourists, residents and investors. During such process, cultural change occurs in terms of reinterpretation of intangible heritage and self-perception. Negative social reactions may occur when local people perceive that the cultural symbols used in bidding and marketing campaigns are not much relevant for the local culture. This situation can lead to detachment from the cultural event as a whole.

This research analyses the inspirations and their inclusion in the projects of two bidding cities: Plymouth, candidate for the title of UK City of Culture 2017, and Malaga, an aspirant for the designation as European Capital of Culture (ECOC) 2016. The two cities focused their cultural strategy in the bidding campaign. In both cases, there was an integration of local and global dimensions in their event themes and discontent raised against how the bidding process was developed. The objective of the research was to analyze political discourses about the festival and social discontent. The methodology consisted of interviews with key informants and analysis of public policy documents, press releases and marketing campaigns.

The increasing importance of bidding for events

Cities have always celebrated festivals and cultural events, but they were mainly for local people. Now, the reasons relate more to improving the image of the city and attract tourists, especially those making short trips out of season and repeating visit. Similarly, improving the image of the city helps to attract investment and, when appropriate economic conditions exist, it helps executives to move to live in the city.

Celebrating festivals and major cultural events has an obvious impact on the economy of cities or regions. Bruno Frey (2000) has pointed out the advantages of this conception of culture on stable programs since it is not required to invest all year in permanent institutions and salaries. Another benefit is the creation of long-term cultural strategies that continue beyond the celebration of the particular event, including an increase in audiences at existing venues in cities. Besides, according to Herrero, Sanz, Bedate and Del Barrio (2012), tourists interested in events spend more than average visitors. Furthermore, cities having few attractions can take advantage of events, thus becoming more competitive.

Richards and Palmer (2010) consider that the extent of the benefits produced by events makes them more attractive than built heritage in cities' cultural and economic strategies. This is because events are more flexible than some types of physical infrastructure, they cost less and have more impact in the short term, and they help to differentiate similar spaces in different cities. It is possible to accept that historic events also have the role of bringing entertainment to tourist attractions and urban spaces contributing to revitalization. According to Bernad Monferrer (2011), events can better provide entertainment and atmosphere and they meet the need for new creative participation of tourists, providing them with multisensory experiences linked to the spaces in which they are developed. Besides, an improvement in networking and co-operation between stakeholders in culture occurs during the celebration of high-rise events.

Celebrating high impact events has also the capacity of encouraging participation in culture, narrowing the division between 'high' and 'low' culture. Similarly, events have a positive impact on internal perceptions, increasing local pride and community spirit. Many of the projects developed in events lasting several weeks help question perceptions of local people and open their minds.

Depending on the international scope and regularity of events, researchers classify them into those, which are part of a regular program, special events and mega-events (Grix, Brannagan and Houlihan 2015; Law 1996; Pereira, Rodrigues and Ben-Akiva 2015). Special events take place annually or only once. Mega-events have an international scope and a large media impact such as the Olympic Games, International Exhibitions, America's Cup and Tall Ships' Races. In the case of mega-events, the enhancement of the city's image occurs just after the presentation of the bid (Giroud and Grésillon 2011) regardless of the final decision (Richards 1999).

In the case of some major international events, the brand value of the event is more powerful than the one of the cities itself. This phenomenon occurs with the designation of ECOC, where the brand has eclipsed individual cities. This happened at least since 1995 because from that date the major European capitals ceased to be elected by the respective governments as candidates (before this date, Athens, Florence, Madrid, Paris and Lisbon had been elected, but celebrations of the ECOC had no remarkable impact in the cultural life of these cities). In other cases, the event brand and the city brand feed off each other. Barcelona 1992 constitutes one of the most successful examplesin this regard (Richards and Wilson 2007).

Culture in all its aspects is the focus of a growing number of events such as the Cultural Olympiad, Capital of Culture with its European, British and American variants, Universal Forum of Cultures, Europride, WOMEX, World Book Capital, World Design Capital and White Night. In Europe, the most desired title is that of ECOC, especially for cities not generally recognized as 'cultural' or undergoing restructuring such as those belonging to the former Soviet bloc (Trócsányi 2011). The main benefits associated with this award are an increase in cultural tourism, improving the image of the city, an acceleration of urban regeneration, the promotion of cultural production and consumption and the promotion in cooperation among administrations, cities and the public and private sectors (Liu 2014).

The increase in the number of cities bidding for the most demanded events contributes to their trivialization due to the multiplication of candidates and selected cities (Muñoz Ramirez 2012). Besides, the competitive advantage these events held in the past is dilutes as benefits are now shared among a growing number of cities (Meethan and Barrera-Fernández 2012). Thus, strategies to improve loyalty and stimulate future visits to the event must be encouraged, combining an enjoyable program in a comfortable environment at a reasonable price (Tanford and Jung 2017).

Some cities have chosen to fill the calendar with events. To achieve that, stakeholders need to coordinate the various programs through a consistent common strategy. In the current context of the growing importance of creativity and intangible aspects for cultural tourism, events become creators of meaning and a renewed image of the city, adding a dynamic component. In the search of new events, traditional or local activities have achieved a renovated role, due to their reinvention to attract a larger and more diverse audience. However, if the priority is only visitor satisfaction, they may lose their authenticity and, therefore, their interest for the local community (Brida, Disegna and Osti 2013).

Negative impacts of bidding for events and social reactions

The enhancement of the city's image during the bidding process forces cities to invest large sums of money during the bidding process, without considering the costs they will have to face if finally selected. This is one of the aspects in the core of the discussions within local artistic communities, provoking great discontent.

Another aspect worthy of consideration is that stakeholders organize some events too fast to include all opinions of social actors and therefore transparency of the process might be questioned (Dredge and Whitford 2011). Moreover, in many cases the bidding effort gives priority to the interest of tourists rather than that of residents, resulting in the construction of attractions and infrastructure not always needed by local people. There are very clear cases of cities hosting large events overlooking what to do with the facilities once the celebration has ended, as the Universal Exhibition of 1992 in Seville

(Reina Fernandez 2012) or Euro 2004 in Portugal (Grande 2012). Besides, the relationship between major events, gentrification and creation of urban spaces designed almost exclusively for consumption need special consideration (Paton, Mooney and McKee 2012). Related to this aspect is the limited impact of special events to encourage social interaction, being more useful in this regard regular activities that require a more consistent and committed citizen involvement (Askins and Pain 2011). This is of particular importance when new events comply with a neoliberal point of view, disregarding cultural traditions and backgrounds of fragile local communities (Higgins-Desbiolles 2018).

In the case of the process to be selected ECOC, some of the explained criticism has tried to be solved, although little evidence relating to social impact has been presented so far. In particular, the so-called 'citizenship dimension' is considered key to achieving success by several commentators (García and Cox 2013). The European Commission (Decision 1622/2006/EC) deemed social implications as the main objective. A priority is to encourage participation and arouse interest from broader audiences than usual. During Rotterdam 2001, Salamanca 2002 and Liverpool 2008, different social agents considered the events as an opportunity to foster public participation in the cultural sector. Cork 2005 went further and acknowledged a tension between the celebration of a singular event and the potential for excluding local people, which resulted in a loss of public support during the celebration (García and Cox 2013).

Another common criticism is the lack of long-term vision in the event legacy. Due to the usual short-term success, the European Commission (Decision 1622/2006/EC) emphasizes the medium- and long-term importance of these interventions. However, the study delivered by Palmer/Rae Associates (2004) showed lack of planning beyond the cultural year, giving several examples such as Stockholm 1998, Bologna 2000, Helsinki 2000, Prague 2000 and Santiago de Compostela 2000 as cities. As a response to this extended practice, Recommendation 6.1 established a reward for bidding cities that have already developed a long-term cultural policy strategy for their city.

Besides, a common practice in cities celebrating events is the mix of objectives that do not fully match the cultural sphere. In many of them, it is common to focus on urban and economic development, thinking more about the impact on tourism and marketing than in the extension of culture to wider audiences. As a result, a large part of investment focuses on urban renovation and development of infrastructures, which generally turns into an increase in tourism (García and Cox 2013).

Moreover, the decision of what kind of events need promotion in a cultural festival tends to be a controversial one. That is the case of many cities focusing on manifestations of high international level, which often results in the exclusion of local people. Examples such as Porto 2001 show that each stakeholder interprets the celebration very differently. In this case, the European Capital acted as a stimulus to an intense process of urban renewal: communications, public health and cultural infrastructure. A renovation process at the old town greatly improved the image of the city and quality of life of the inhabitants. The increase in tourism has also been considerable.

Alvaro Domingues and João Teixeira, who were involved in drafting the proposal for Porto 2001, explain that designing a program to attract audiences normally excluded was crucial, as well as encouraging public participation. Concerning the topic of audiences, João Teixeira indicates that the program of Porto 2001 acted at two levels: first, to attract new viewers and culture consumers (especially reaching segments of the public which are traditionally excluded or more distanced); and second, to encourage the participation of individuals and groups in the scheduled activities (Teixeira, personal communication, 05.05.2014). In this respect, relations were established with neighbourhood associations, schools and other communities through specific projects. The results were very positive, including initiatives such as the performance of the 1034 *Wodzzeck* opera by Opera Birmingham in the neighbourhood of Aldoar, which implied a strong involvement of residents in one of the most deprived areas of the city. This kind of actions stressed social inclusion so that there was a balance between performances of high culture for a select audience and a wide audience normally excluded from this kind of events (Rodríguez 2014). Nevertheless, the perception among artists and managers was not so positive: many felt that there had been a lack of communication between the organizers and local cultural artists. Great exhibitions and shows took place for large audiences but with little participation from local artists.

The audience welcomed the celebration of the ECOC Porto 2001 because of the quantity and quality of the events while a considerable segment of the artistic community shared their euphoria. However, the change of local government meant a sharp cut in funding and deterioration of official support. By the end of 2001, culture in the city already had a different character. According to artist Eduardo Matos, the new mayor destroyed achievements by Porto 2001 in his first six months, "the city's cultural life was in deep decline. For those who produce and think about art every day it was urgent to do new things. We couldn't hang around waiting" (Gomes 2007: 16).

Another perspective about events in cities relates to rapid sociocultural change. In some way, bidding implies creating a city brand usually based on local identity (Dinnie 2011). From a cognitive anthropology point of view, culture is defined as mental content (Brown 2006; D'Andrade 1995; Goodenough 1964). Therefore, the imaginary realm related to self-perception implies a constant use and reuse of actual or fictional concepts, historic events and symbols during the branding process. As a result, events can either add strength to the local identity or lead to intangible heritage loss due to banalization. The whole topic becomes relevant since culture constitutes one of the pillars of sustainability (Burford et al. 2013). In this sense, it is possible to question all of the economic goals described earlier in some cases where cultural sustainability has a negative impact, especially for locals' identity or intangible heritage.

Impact of the bidding process and reactions in Plymouth

Plymouth possesses extensive experience in organizing events related to the sea, at least since the 1820s. Major regattas that take place regularly in the city include Port of Plymouth Regatta and Plymouth Classic Boat Rally hosted every year, Fastnet Race hosted every two years, and the Two-Handed Transatlantic Race and Artemis Transat hosted every four years.

Among all the events held recently, the most impacting one has been the America's Cup World Series Yacht Racing 2011. It was highly positive in terms of visitor numbers, especially because of the attraction of international tourists, and had an extraordinary impact concerning the city's media projection. Since the success of this race, celebrating events has become a priority in Plymouth, being the nearest target to celebrate the 400th anniversary of the voyage of the Mayflower in 2020.

The relationship of Plymouth with the sea is not only restricted to events. The city is seeking to become recognized as the British city of the sea, as reflected in all economic and tourism plans. The new city brand is 'Plymouth, Britain's Ocean City', which celebrates the importance of the city as a port of travel of global significance. Plymouth was the departing point of the voyages of James Cook, Charles Darwin, Francis Drake, Robert Falcon Scott and the Pilgrim Fathers. Plymouth also wants to emphasize its global links, reflected in the 51 homonymous cities that exist worldwide. The new brand also reflects the aspirations of Plymouth to become a city that inspires dynamism, depth and breadth of vision, ideas suggested by the word 'ocean' (Jones 2013). Moreover, the choice of the sea as thematic line relates to the local economic base, with an important presence of shipbuilding, ship repair, biological research, aquaculture, biotechnology, fishing and water sports.

As a result, it is not surprising that the city aspired to become UK City of Culture 2017 focusing in sea-related topics. The core idea was to express how the ocean defines the city through its history, culture, economy and society. Among the main characters related to the sea mentioned by the proposal were Francis Drake, James Cook and Charles Darwin. The application also highlighted the role of the city in national marine defence, since it hosts Devonport Naval Base, the largest naval base in Western Europe (Royal Navy 2012).

The announcement of Plymouth's bid was a strategic engine to reactivate the local economy. Abby Johnson, executive director of the Plymouth Culture Board, stressed the opportunity to the city to host events of great economic and media importance, citing examples such as the Irish city of Derry/Londonderry, Turner Prize, BBC Sports Personality of the Year, Brit Awards and the UK Film Festival. In the words of Johnson (2012: 25) "Culture is a mechanism to deliver wider economic and social goals, it is a catalyst for change".

In the end, Hull became UK City of Culture 2017. According to Calvin Watts (2013), the election of this city, in particular, was as a loan, or an opportunity that the city should be able to take advantage. Hull had recently suffered a severe economic crisis following the collapse of the fishing industry and government austerity measures. Holding the title was an incentive to regenerate the local economy, especially in areas related to culture and tourism.

A useful conclusion drawn from the election of Hull is the strength brought by a strong link between previous achievements and future aspirations. Watts explains the election of Hull by the quality of the submitted project, which reflected the unique and happy spirit of the city and is a continuation of cultural events from previous years such as Freedom, Humber Mouth Literature Festival and Hull and Humber Sesh Jazz Festival (Watts 2013). The bid organizers took advantage of these dynamics, presenting the proposal as a natural continuation of a cultural impulse of great popular acceptance. Nevertheless, the success of Hull will need further evaluation in the following years after the celebration of the event. Sustained participation, reinvestment, an increase in the number of visitors, an extension of the tourist season and the creation of new jobs in the sector should be priorities of the legacy of the event.

The selection of Hull did not affect the momentum of Plymouth to host major events. City officials kept the focus on the revival of the city from culture and in July 2013, an open working meeting was organised to rethink projects and lines of action for the future. It was, in the words of Mayor Tudor Evans, to define something as important as the cultural future of the city. However, the key point is that the main promoters of this strategy clearly understood since the beginning that the sense of culture employed should be inclusive and not be identified with 'high culture' in the traditional way (Eve 2013). The specific target of the city in the cultural field is the celebration in 2020 of the 400th anniversary of the departure of the Pilgrim Fathers on board the Mayflower, presumably producing an impact in both the United Kingdom and United States. The US president and the head of the British state in the presence of all American presidents will address a speech on the Hoe. Besides, the city will build a new memorial, they will launch a new race between Plymouth (England) and Plymouth (Massachusetts) and the installation of a replica of the Mayflower in Sutton Harbour is under discussion. The promotion of American tourism is also behind this major event. In particular, there is a commitment to improving city break tourism and creating an attractive product for the American market: the so-called American Heritage Trail (Vinken 2013).

Impact of the bidding process and reactions in Malaga

Malaga has no tradition of hosting large singular events. Being selected ECOC 2016 has been a central objective in the city. The bidding process coincided with several projects to attract a larger number of urban cultural tourists, maintain a relatively improvised profile and lacked participation.

The city presented its project in Madrid, in September 2010, along with 14 other candidates. After the rejection, the foundation in charge dismantled the bidding process, showing the lack of long-term vision of the cultural project if we compare it with the previous case. Stakeholders did not achieve a common vision and they did not support enough local cultural and artistic initiatives, which is essential in the evaluation of candidates.

One of the main problems was the reduction of the cultural project to the construction of new museums, especially local branches of international icons. This process began discreetly with the opening of Museo Picasso, taking advantage of the fact that the painter was born in the city, thus establishing a successful co-branding strategy with Picasso. The co-branding strategy was designed by public authorities and implied not only the opening of the Museum but also a deep renovation of the streets surrounding it, the promotion of souvenirs related to the artist, new street signage and a variety of tourist promotion campaigns (Barrera-Fernández 2013). This intense program of cultural marketing centred in the figure of the Spanish artist was the inspiration in 2012 of a work of art: *Almost all of Picasso*, an ironic commentary from the artist Rogelio López Cuenca, which took the form of an archive of images and souvenirs. Going beyond this first step, in recent years a number of franchises have been boosted with the Museo Carmen Thyssen, Pompidou Centre and Museo Estatal de San Petersburgo.

The case of Malaga and the figure of Picasso have become relevant in terms of cultural sustainability. Indeed, the painter was born in the city but left it when he was nine years old. However, historic facts reveal that his actual link to the city was weak ('Laniado vincula' 2013). In some way, an idealization process has led to historic distortion as it has happened in other historic cases (Sommer 1991). Locals remain faintly related to this figure, despite the great effort to create this relationship. However, other more traditional items of local culture such as its relationship with the sea or historically relevant characters are increasingly lost.

The focus on contemporary art was crucial in the proposal to be competitive with other candidate cities such as Cordoba, which rooted its project in its rich history and renowned heritage. Thus, Malaga strongly promoted the relationship of the city with Picasso. A consequence of this focus on the contemporary was the lack of interest in numerous heritage assets that do not fit with the discourse of avant-garde and spectacularity. As a result, *façadism* and neglect have become a common practice in the treatment of built heritage in the historic city.

The project finally presented was entitled 'Infinite City'; it had seven strategic lines:

1 Paradise city: It emphasized the presence in the city of artists and intellectuals from around the world and its cosmopolitan character.

2   Ciudad jonda: It addressed the culture as improvisation and spontaneity.
3   Building gardens: It highlighted the contrast between the city's green areas and the urban problems of the Costa del Sol.
4   Tradition and future: It emphasized the achievements of the city when it looked to the future, citing as examples the Phoenician period and the urban boom of the 1960s.
5   In danger of freedom: It focused on culture as a debate.
6   Desire Caught by the Tail: It intended to accommodate art of the highest level.
7   Prodigious city: It focused on contemporary approaches of bullfighting, Easter, the night of San Juan, flamenco music and other traditions (Malaga City Council 2010).

Significantly, the project emphasized the opportunity that culture could bring as a way to regenerate empty spaces and run-down areas through the construction of cultural facilities. On one side this strategy showed the focus of the project on new cultural facilities instead of promoting existing ones, and on the other side, it disregarded large areas of the historic city that do not fit in the image that is promoted to tourists; for them, there has not been any alternative than a slow process of neglect.

In general, the bidding document reduced culture to the spectacle. Its objectives vanished after not been selected, showing their lack of soundness, the absence of public participation and the improvisation in the whole process. Currently, there are no significant events planned, and among those that exist, it is difficult to draw a shared vision.

During the bidding process, voices against the project arose, arguing that it was contradictory to promote official culture while trying to dismantle popular initiatives and independent artistic movements. It was the case of the Neighborhood Theatre Festival. Another controversial decision was the eviction of the Casa Invisible, an active social centre and producer of independent culture. The bidding process coincided with the years when the city representatives decided to turn the city centre into a thematic stage to suit visitors' expectations, and to deliver the explained strategy of opening museum branches external from the local culture.

The debate of the local cultural policy is far from reaching an end. Carlos Taillefer, producer and filmmaker, has contrasted the model of the eighties and nineties of a city promoting good quality classical music and orchestra concerts with that of recent years when the priority has become to open museums of commodified culture, neglecting certain types of programs such as theatre and music in favour of franchise museums (Zotano 2014). The process disregarded local artists. As the poet Alejandro Simon Partal points out,

> a cultural boom has more to do with the artists than with the institutions. I guess, from a distance, that there are very good artists, writers,

poets and musicians in Malaga, and public and private institutions are realizing that culture equals leisure equals tourism equals money.

(Zotano 2014)

Conclusions

In the two cities researched, it is possible to observe the integration of both local and global dimensions in their event themes. Plymouth focused on its physical, historic and economic connections with the ocean, reconceptualising it as a means of embracing ideas like freedom, open horizons and inspiration. Malaga adopted the same process, in this case rooting the project in the figure of Pablo Picasso, connected to values such as avant-garde and innovation. The bidding process has served to embed new ideas in the cities' imaginary, affecting tourism promotion, cultural policy and events strategy. However, there is a significant difference between both cases. Plymouth has kept on bidding for events related to the sea since this is something deeply rooted in local identity and shared by its residents. On the other hand, Malaga put the focus on the idea of culture as spectacle and brand, lacking root in local cultural movements leading to abandonment in the end.

Public reactions to the bidding process were diverse in Malaga and Plymouth, but they share in common that different sectors participated in the debate, from local artists to the economic sector, and their response was sometimes opposite. The tourist industry showed enthusiasm, while the cultural sector expressed from indifference to rejection, especially notable in the case of Malaga. If we compare these results to those from European Capitals of Culture, we can notice that there were clear economic and cultural benefits, especially related to the association of culture with spectacle. However, positive impacts at a social level or for local cultural producers remain disregarded. A specific topic for further research is the sociocultural transformation and whether the impact results beneficial or not in terms of sustainability.

These experiences can be useful for similar cases in other countries. Practical implications for destination managers might include taking into account the opinion of local social actors and whether the impact of the event is seen as positive or not. Events imply cultural transformations, especially regarding identity and intangible heritage. Economical perspective cannot be the only one involved. The impacts of events in social sustainability might as well become a theoretical realm and need more profound research.

References

Askins, K. and Pain, R. (2011) 'Contact zones: participation, materiality, and the messiness of interaction', *Environment and Planning D: Society and Space* 29(5): 803–821.

Barrera-Fernández, D. (2013) 'La tematización de la ciudad en torno a un personaje. El caso de Picasso y Málaga', *AGIR – Revista Interdisciplinar de Ciencias Sociais e Humanas* 1(5): 8–20.

Bernad Monferrer, E. (2011) 'Eventos y ciudad: los eventos como elementos clave para la proyección territorial', *Actas ICONO 14 - nº 8. II Congreso de Ciudades Creativas*, Madrid, 26–28 October 2011.

Brida, J.G., Disegna, M. and Osti, L. (2013) 'Perceptions of authenticity of cultural events: a host-tourist analysis', *Tourism, Culture & Communication* 12: 85–96.

Brown, P. (2006) 'Cognitive anthropology', in C. Jourdan and K. Tuite (Eds.), *Language, Culture, and Society: Key Topics in Linguistic Anthropology*. New York: Cambridge University Press, pp. 96–114.

Burford, G., Hoover, E., Velasco, I., Janousková, S., Jimenez, A., Piggot, G. and Harder, M.K. (2013) 'Bringing the "missing pillar" into sustainable development goals: towards intersubjective values-based indicators', *Sustainability* 5(7): 3035–3059.

D'Andrade, R.G. (1995) *The Development of Cognitive Anthropology*. Cambridge: Cambridge University Press.

Decision 1622/2006/EC of the European Parliament and of the Council of 24 October 2006 establishing a Community action for the European Capital of Culture event for the years 2007 to 2019.

Dinnie, K. (2011) *City Branding: Theory and Cases*. Palgrave Macmillan.

Dredge, D. and Whitford, M. (2011) 'Event tourism governance and the public sphere', *Journal of Sustainable Tourism* 19(4–5): 479–499.

Eve, C. (2013) 'So, we lost the UK City of Culture 2017 bid... but what are we going to do now?', *Plymouth Herald*, 28 June 2013.

Frey, B. (2000) *La economía del arte*. Barcelona: La Caixa-Servicio de Estudios.

García, B. and Cox, T. (2013) *European Capitals of Culture: Success Strategies and Long-Term Effects*. European Parliament.

Giroud, M. and Grésillon, B. (2011) 'Devenir capitale européenne de la culture: principes, enjeux et nouvelle donne concurrentielle', *Cahiers de géographie du Québec* 55(155): 237–253.

Gomes, K. (2007) 'Porto off', *Público*, 17 August 2007.

Goodenough, W.H. (1964) 'Componential analysis of Konkama Lapp kinship terminology', in Goodenough, Ward (Ed.), *Explorations in Cultural Anthropology*. New York: McGraw-Hill, pp. 221–238.

Grande, N. (2012) 'Portugal eventual: de Lisboa'94 a la Eurocopa 2004. Legado de un decenio de grandes eventos urbanos', *Seminario Internacional sobre Eventos Mundiales y Cambio Urbano*, Seville, 26–28 November 2012.

Grix, J., Brannagan, P.M. and Houlihan, B. (2015) 'Interrogating states' soft power strategies: a case study of sports mega-events in Brazil and the UK', *Global Society* 29(3): 463–479.

Herrero, L.C., Sanz, J.A., Bedate, A. and Del Barrio, M.J. (2012) 'Who pays more for a cultural festival, tourists or locals? A certainty analysis of a contingent valuation application', *International Journal of Tourism Research* 14: 495–512.

Higgins-Desbiolles, F. (2018) 'Event tourism and event imposition: a critical case study from Kangaroo Island, South Australia', *Tourism Management* 64: 73–86.

Johnson (2012) Conference on "the future of our heritage" in Plymouth Guildhall. 'Plymouth set to bid for UK City of Culture title', *Western Morning News*, 30 October 2012.

Jones, P. (2013) 'Are we ready for 2020?', *Tourism and Visitor Economy Conference*, Plymouth, 12 September 2013.

'Laniado vincula el rechazo a Picasso en Málaga con la creación de su museo en Barcelona' (2013) *Diario Sur*, 2 August 2013.

Law, C.M. (1996) *Urban Tourism. Attracting Visitors to Large Cities.* London, New York: Mansell Publishing Limited.

Liu, Y. (2014) 'Cultural events and cultural tourism development: lessons from the European Capitals of Culture', *European Planning Studies* 22(3): 1–17.

Malaga City Council (2010) *El proyecto de Málaga 2016 se basa en siete líneas programáticas que aglutinan todas las artes'.* Málaga: Fundación Málaga Ciudad Cultural.

Meethan, K. and Barrera-Fernández, D. (2012) 'Urban transformations from being designated European Capital of Culture', *Seminario Internacional sobre Eventos Mundiales y Cambio Urbano,* Seville, 26–28 November 2012.

Muñoz Ramírez, F. (2012) 'Los megaeventos en la ciudad del siglo XXI: cuatro hipótesis para el futuro del acontecimiento urbano', *Seminario Internacional sobre Eventos Mundiales y Cambio Urbano,* Seville, 26–28 November 2012.

Palmer, R. (dir.). (2004) *Report on European Cities and Capitals of Culture - Part II.* Brussels: Palmer/RAE Associates.

Paton, K., Mooney, G. and McKee, K. (2012) 'Class, citizenship and regeneration: Glasgow and the Commonwealth Games 2014', *Antipode* 44: 1470–1489.

Pereira, F.C., Rodrigues, F. and Ben-Akiva, M. (2015) 'Using data from the web to predict public transport arrivals under special events scenarios', *Journal of Intelligent Transportation Systems* 19(3): 273–288.

Reina Fernández, J.C. (2012) 'Eventos mundiales: teatro versus realidad. La relevancia de los espacios públicos urbanos', *Seminario Internacional sobre Eventos Mundiales y Cambio Urbano,* Seville, 26–28 November 2012.

Richards, G. (1999) 'The European cultural capital event: strategic weapon in the cultural arms race?', *Journal of Cultural Policy* 6(2): 159–181.

Richards, G. and Palmer, R. (2010) *Eventful Cities. Cultural Management and Urban Revitalization.* Amsterdam: Elsevier.

Richards, G. and Wilson, J. (2007) *Tourism, Creativity and Development.* London: Routledge.

Rodríguez, M.A. (2014) 'Santiago de Compostela and Porto. Two European Cities of Culture between spectacle and crisis', *8th International Conference on Cultural Policy Research,* Hildesheim, 9–12 September 2014.

Royal Navy. (2012). [Internet]. London: Secretary of State for Defence; 2012 [cited 2019 Aug 27]. Devonport Naval Base. Available from: http://www.royalnavy.mod.uk/The-Fleet/Naval-Bases/Devonport

Sommer, D. (1991) *Foundational Fictions: The National Romances of Latin America,* Vol. 8. University of California Press.

Tanford, S. and Jung, S. (2017) 'Festival attributes and perceptions: a meta-analysis of relationships with satisfaction and loyalty', *Tourism Management* 61: 209–220.

Trócsányi, A. (2011) 'The spatial implications of urban renewal carried out by the ECC programs in Pécs', *Hungarian Geographical Bulletin* 60(3): 261–284.

Vinken, A. (2013) 'Are we ready for 2020?', *Tourism and Visitor Economy Conference,* Plymouth, 12 September 2013.

Watts, C. (2013) 'City of culture is not the answer to all Hull's problems – it's a catalyst', *The Guardian,* 25 November 2013.

Zotano, J. (2014) [Internet]. 'La explosión cultural de la ciudad desde la distancia' [cited 2019 Aug 27], *La Opinión de Málaga,* 21 December 2014. Available from: https://www.laopiniondemalaga.es/cultura-espectaculos/2014/12/21/explosion-cultural-ciudad-distancia/731386.html

12 MICE tourism development in Ethiopia

Gebeyaw Ambelu Degarege and Brent Lovelock

Introduction

The MICE (Meetings, Incentives, Conferences and Exhibitions) sector (Dwyer & Forsyth, 1997) is one of the more rapidly growing market segments of the tourism industry (Jones & Li, 2015; Mistilis & Dwyer, 2000; Spiller, 2002; Weber & Chon, 2002). According to the International Meetings Statistics Report of the Union of International Associations (UIA), in 2016, there were 468,700 international meetings held worldwide (UIA, 2017). Parallel to its growth, the MICE sector has received increasing attention from the public and private sectors and is seen as a critical area of growth for tourism destinations (Mistilis & Dwyer, 2000; Oppermann & Chon, 1997; Weber & Roehl, 2001). It is now common to see local and national tourism strategies that position events as central components for destinations (Getz, 2009), as a form of attraction, as development catalysts, and for destination branding and image-making (Getz, 2008). This chapter is prompted by the growth in the MICE sector and the relative paucity of attention given to the challenges of developing the MICE sector within developing world settings (Rogerson, 2005). With these concerns in mind, this chapter analyses the current position of and challenges for the MICE tourism development in the context of Addis Ababa, Ethiopia.

Addis Ababa is the capital city of Ethiopia and the headquarters of the African Union (AU), the United Nations Economic Commission for Africa (UNECA) and the seat of various multilaterally and bilaterally accredited missions, delegations and institutions. Consequently, Addis Ababa is often identified as one of the most significant diplomatic centres in the world, after New York and Geneva (Elias, 2016), and the city currently serves as a frequent venue for international conferences and business events. Being a regional and international hub for many meetings, the challenges facing Addis Ababa regarding growing its MICE tourism are representative of those facing many rapidly developing African nations, which include infrastructure, destination image, governance and political problems.

This chapter begins by discussing the development of the tourism sector in Ethiopia, then focusing on Addis Ababa's MICE sector, providing an analysis of the historical and contemporary trends. The next section discusses the potential for growth of the MICE sector in Addis Ababa and identifies opportunities for MICE tourism development. This chapter then discusses the most prominent strategic issues and challenges in relation to increasing the performance of the MICE sector and suggests mechanisms to harness the existing potential of the city and improve the contribution to the welfare of its residents and the business community.

Tourism sector development in Ethiopia

Ethiopia, the home of the earliest hominids on earth (Johanson & Taieb, 1976; White, Suwa, & Asfaw, 1994) and with a recorded history that dates back more than 3,000 years, is one of the world's oldest civilisations (Phillipson, 1998; Selassie, 1972). In recent years, Ethiopia has become one of the fastest growing economies in the world. The country has registered a double-digit average real GDP growth of 11.4% over the period 2003/4–2010/11 (MoFED, 2014a). Despite such an ancient history and remarkable current growth, Ethiopia is still one of the poorest countries in the world. Its per capita income of $660 is lower than the regional average in sub-Saharan Africa (World Bank, 2017). Until recently, Ethiopia's development policy was heavily informed by agriculture and rural development under the umbrella of the Agricultural Development-led Industrialization (ADLI) programme. This exclusively two sector centred development orientation can be considered as one of the critical reasons for the current low level of tourism development in the nation (Degarege & Lovelock, 2019). The relative significance of tourism has been recognised over the last two decades, and tourism has been embraced in Ethiopia's development strategies, i.e. the Plan for Accelerated and Sustained Development to End Poverty (PASDEP) (MoFED, 2006), Millennium Development Goals (MDGs) and the Growth and Transformation Plan (GTP) I & II (MOFED, 2014b).

Historically, tourism in Ethiopia is a phenomenon of the 1960s, where the first institutionalised development of tourism was undertaken in 1962 with the formation of the Ethiopian Tourist Office (Sisay, 2009). Since then the role of tourism has varied depending on the strategic goals, as well as the ideological positions of the various ruling regimes. The progress in international tourism growth during the imperial regime declined during the military government from 1974 to 1991 (Degarege & Lovelock, 2019). This is because of the adverse effects of measures taken to safeguard the Socialist Revolution from external counters of tourist generating countries, political instabilities and restrictions on entry and free movement of tourists (Sisay, 2009). It is only very recently that the country has experienced growth in international tourism, both in arrivals and in tourism receipts and there remains further untapped potential. The National Tourism Development Policy (NTDP) document of Ethiopia adopted in 2009 specifies that "enhancing the development impacts of tourism by properly developing and utilizing the tourism potential with which the country is endowed is a matter deserving focus" (Federal Democratic Republic of Ethiopia, Ministry of Culture and Tourism (MoCT), 2009, p. 38). To optimize the development impact of tourism, reforms of institutional and governance structures have taken place. The most prominent was the upgrading of tourism sector governance from a Tourism Commission into a fully-fledged ministry in the establishment of the MoCT in 2005. In addition, the formulation of the Tourism Transformation Council (TTC) and the Ethiopian Tourism Organization (ETO) in 2014 reflects an increased recognition of the role of tourism (Degarege & Lovelock, 2019). Currently, an objective of the government is to become one of the top five tourist destinations in Africa by 2025 through the development of Ethiopia's cultural wealth and national attractions (MoCT, 2012). With the intention of developing the sector and achieving this goal, Ethiopia has created a worldwide tourism marketing slogan "Ethiopia, Land of Origins" that emphasizes the country's ancient and rich history, the origin of humankind, the birthplace of Arabica coffee and the source of the Blue Nile River (ETO, 2017).

According to the World Economic Forum (WEF), Ethiopia's global Travel and Tourism Competitiveness Index (TTCI) ranking has shown continuous improvement over the years, rising from position 123/133 in 2009 (WEF, 2009) to 116/136 in 2017 (WEF, 2017). The TTCI for 2017 shows that out of 136 countries, Ethiopia performs relatively well in some areas including environmental sustainability (position 56), price competitiveness in travel and tourism sector (position 64) and natural resources (position 69). However, the rankings for tourist service infrastructure (position 129), human resource and labour market (position 125) and ICT readiness (position 125) are the worst performing areas (WEF, 2017) that the country will need to address if its tourism potential is to be harnessed.

The level and contribution of tourism to Ethiopia's economy are still small but showing signs of growth (see Figure 12.1). In the year 1963, the international tourist arrivals were just 19,215. As shown in Figure 12.1, since the 2000s, international tourist arrivals have grown from 135,954 in 1990 to 870,597 in 2016. In 2016, international tourist arrivals grew by 0.79%, an increase of 6,855 tourists over the previous year. The number of international tourist arrivals has increased by an average annual growth rate of 13.83% over the period 2000–2016. This has been matched by growth in total international tourism receipts, which for the same period grew by an average of 39.57% per year to reach a total of Ethiopian Birr (ETB)[1] 3,259,515,168.00 in 2016 (MoCT, 2016). In 2016, the direct and total contribution of tourism to Ethiopia's GDP were USD 1,402.5mn (2.2%) and 3,620.6mn (5.7%), respectively. In the same year, tourism supported 1.9% of direct employment and 5.1% of total employment. Regarding foreign exchange, in 2016, Ethiopia generated USD 1,073.3mn in visitor exports, which accounts for 19% of the total exports in 2016 (WTTC, 2017).

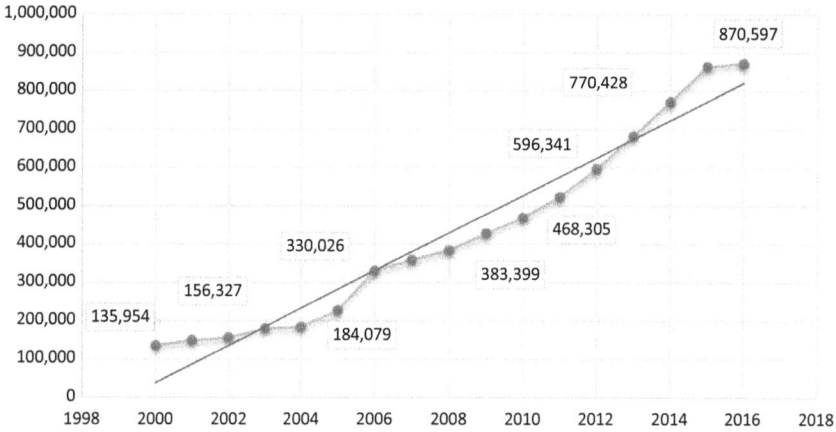

Figure 12.1 International tourist arrivals, 2000–2016.
Source: MoCT (2016).

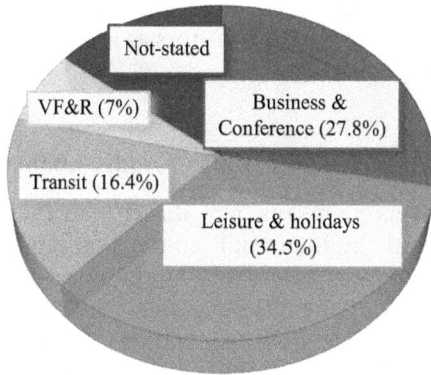

Figure 12.2 International tourism by the purpose of visit, 2010–2015.
Source: MoCT (2016).

The international tourist arrivals by purpose of visit show that Ethiopia's tourism sector is a composition of leisure and holidays, conference and business, transit, and visiting friends and relatives. Figure 12.2 shows that travel for holidays, recreation and other forms of leisure accounted for 34.5% of international tourist arrivals for the period 2010–2015. Some 27.9% of all international tourists reported travelling for business (18.7%) and conference (9.2%) purposes. About 16.4% of the international visitors arrived for the purpose of transit, and another 7% travelled for visiting friends and

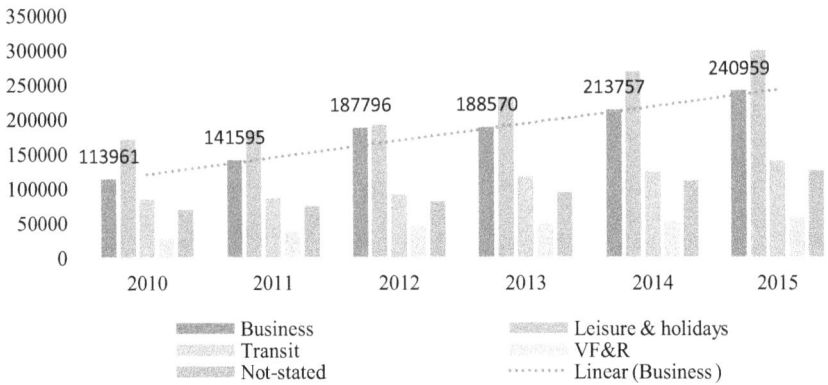

Figure 12.3 Growth and contribution of the business purpose of visitors, 2010–2015.
Source: MoCT (2016).

relatives (MoCT, 2016). As shown in Figure 12.3, the share of international visitors for conferences and business purposes grew from 113,961 in 2010 to 240,959 in 2015 – more than doubled (MoCT, 2016). These figures collectively underline business and MICE visitors as a significant market segment of Ethiopian tourism.

Situating the MICE sector potentials of Addis Ababa

Addis Ababa whose name means 'new flower' was established in 1886. It is the largest city and the industrial, commercial and cultural centre of the country. Its economy is dominated by the services sector, which contributes to about 77% of the city's economy, followed by the industrial sector that contributes 22% (United Nations Human Settlements Programme (UNHSP), 2017). Addis Ababa is primarily dependent on business travel (Addis Ababa Culture and Tourism Bureau (ACTB), 2014; MoCT, 2015). The vision of the Addis Ababa City Culture and Tourism Bureau is "making the city one of the top 5 Favourite Tourist Destination and Cultural Cities in Africa by 2020 through creating a society that is supportive, beneficiary and responsible for its culture, nature and historical treasures" (ACTB, 2014, p. 13). Based on their study on private sector and community engagement in Ethiopia, Mitchell and Coles (2009) suggest that the MICE sector may have a considerable economic impact, possibly even greater than that of leisure tourism sector in Addis Ababa. Since business tourism is a dominant activity, Addis Ababa has attracted the lion's share of visitor expenditure in Ethiopia, accounting for almost 40% of all visitor expenditure in 2015 (MoCT, 2015). Similar to the national tourist arrival growth, the international tourist arrival in Addis Ababa for the period

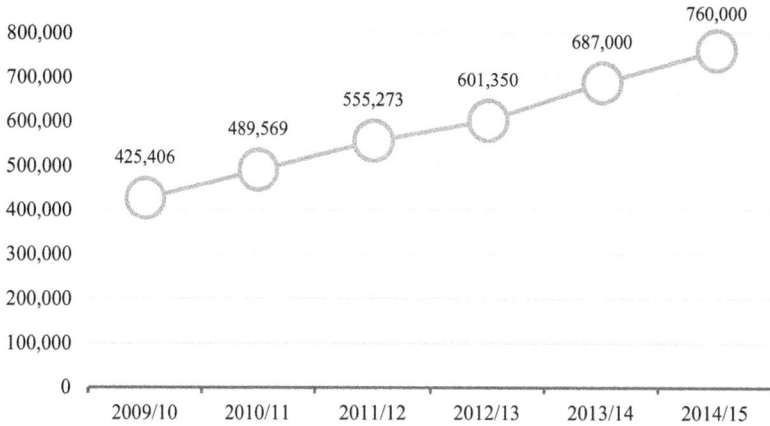

Figure 12.4 International tourist arrivals in Addis Ababa, 2009/2010–2014/2015.
Source: ACTB (2014).

2009/2010–2014/2015 showed growth from 425,406 to 760,000. For the same period, tourism receipts grew from ETB[2] 2.2 billion to 96.5 billion (ACTB, 2014; Figure 12.4).

However, like most African countries' marketing and destination development, most initiatives are targeted at leisure tourism and overlook the importance and potential of this lucrative segment (Rogerson, 2015). The Addis Ababa Tourism Marketing Strategy (2016–2020) currently implemented through the Addis Ababa City Administration and coordinated by the city's Culture and Tourism Bureau underlined the need for going beyond a narrow focus on leisure-based tourism. The strategy capitalises on the city's potential for MICE tourism development and indicates the apparent desire to develop the sector further (MoCT, 2015). This recognition of the MICE sector relates to the growing political and economic significance of Addis Ababa that is catalysing rapid and new investment in high-end and middle-level accommodation units and convention centres. According to the 2016 Global Cities outlook[3] report, Addis Ababa's rank was 114 in 2016, compared with 121 in 2015 (Kearney, 2016). The same report using the Global Cities Index[4] ranking scheme ranked Addis Ababa 100 in 2016 and 85 in 2015 (Kearney, 2016). As shown in Table 12.1, based on the number of international meetings held in 2015, the International Congress and Convention Association (ICCA) ranked Addis Ababa seventh in Africa and 216 globally in 2015. Ethiopia was ranked seventh in Africa and 79th globally based on the number of meetings held in 2015 (ICCA, 2017). Overall, considering the information from earlier rankings, there is a need to work more on making Addis Ababa suitable for international organisations (Table 12.2).

Table 12.1 MICE destinations in Africa, 2016

Country	Number of meetings	Africa ranking	Global ranking
South Africa	125	1	34
Morocco	37	2	59
Egypt	18	3	71
Kenya	18	3	71
Rwanda	18	3	71
Ghana	13	6	77
Ethiopia	12	7	79
Nigeria	11	8	81
Tanzania	10	9	84
Uganda	10	9	84

Source: ICCA (2016).

Table 12.2 MICE cities in Africa, 2016

City	Number of meetings	Africa ranking	Global ranking
Cape Town	62	1	39
Marrakech	19	2	134
Kigali	18	3	144
Durban	16	4	160
Johannesburg	14	5	186
Nairobi	13	6	203
Addis Ababa	12	7	216
Stellenbosch	12	7	216
Accra	11	9	239
Cairo	10	10	256

Source: ICCA (2016).

The supply-side in the MICE sector includes professional event managers, destination management organizations, the MICE event venues and a multitude of event suppliers that play essential roles in MICE events (Locke, 2010). Core conference and convention facilities and infrastructure need the support of adequate access and communication facilities such as phone and email, to name a few essential components. As mentioned earlier in this chapter, Ethiopia lags behind in terms of tourist service infrastructure, human resource and labour market, and ICT readiness. Moreover, a robust and efficient MICE tourism development strongly relies on the quality of the workforce (McCabe, 2012), which is on the frontline of the sector and responsible for ensuring international standards. Accordingly, MICE event quality depends on the performance of every single service that formulates the most significant event tourism system (Getz, O'Neill, & Carlsen, 2001). The following section addresses the key supply-side components for MICE tourism in Addis Ababa and situates its potential.

Political role and presence of international organizations

The early development of the meeting and convention industry in Addis Ababa is not well documented. This might be a result of the low attention given to the sector. The city's MICE tourism potential is highly related to the political role of the nation and the presence of international organisations, and the recent economic growth and relatively open door diplomatic and trade policy that leads to the influx of international business and diplomatic visitors. Ethiopia is one of the founding members of the United Nations (UN),[5] and the AU, formerly known as the Organization of African Unity (OAU). It is one of the political and diplomatic capitals in the world currently serving as the headquarters of the UNECA since 1958, the AU since 1963, the UN Regional Bureau, the African Standby Force, the Pan African Chamber of Commerce and Industry (PACCI) and a large number of foreign diplomatic missions. Currently, more than 92 embassies and consular representatives, and many global Non-Governmental Organisations (NGOs) are based in the city. Moreover, while there is a well-established relationship between political events and tourism, the meetings, conventions and conferences are held under the auspice of the parent companies, i.e. government organisations, UNECA, AU, and the planning and management of such events as a tourism undertaking have not materialised. Consequently, Addis Ababa is yet not adequately benefiting from the conferences and meetings (ACTB, 2014). This is mainly because of the lack of a MICE strategy that harnesses the existing potential into real benefits.

The current state of convention and meeting centres

The basic and primary facilities and infrastructure for MICE tourism are the meeting places that include convention centres, conference rooms and halls, trade fairs and exhibition centres (Sylla, Chruściński, Drużyńska, Płóciennik, & Osak, 2015). The formal establishment of international meetings and convention centres purposively built to provide facilities for conferences, exhibitions and events is related to the establishment of the OAU Hall that was inaugurated in 1961. The new AU headquarters funded by China became operational in 2012, and numerous events have been hosted. The UNECA accommodates 1,500 participants at a time; second is the AU hall with a 657-seat capacity. In addition to such intergovernmental convention centres, an independent state-owned centre named the Addis Ababa Exhibition Center was established in 1983. In general, private sector involvement in convention centres is quite limited. A privately owned convention centre, the Millennium Hall, was opened for celebrations of the Ethiopian calendar year 2000 in 2008, while the latest to open is the Addis-Africa International Convention and Exhibition Center (AAICEC), which will host up to 35,000 people in 14 separate parallel sessions (AAICEC, 2017). Conducting an inventory and developing a database of MICE venues and services and facilities are essential for successful planning and management of the sector. At the same time, mechanisms to harness the development potential of small-scale events need to be considered.

Private sector involvement in the event sector

The MICE sector is essentially public sector led and private sector driven in Abbas Ababa. While a multitude of domestic and international events have been organized and hosted, i.e. festivals, sports (e.g. Ethiopian Great Run), religious events (e.g. *Meskel*/finding of the true cross, *Timket*/epiphany and *Gena*/Christmas), exhibitions, trade fairs and conferences, organizing an event of any kind as a private business is a relatively recent phenomenon. While yet in a low state of development, the incremental growth in private sector involvement is visible. Currently, there are about 18 event organizing private enterprises operating in Addis Ababa (MoCT, 2016). Most of the event organizers work on small-scale local events such as weddings and music festivals. Vibrant and internationally competent private meeting planners and event organizers are yet to be established in the country. This is evidently happening because of the lack of precise event planning and policy that specifically guides the event sector development. Moreover, government incentives are essential to foster private sector engagement in the event business.

Accommodation sector development

Following recent growth and performance in the MICE sector, as well as potential for growth, Addis Ababa has started experiencing a flourishing accommodation sector. To support growth in the tourism sector, the government provides incentives such as duty-free imports of hotel furniture and loans for the construction of star-rated hotels (World Bank & MoCT, 2012). Currently, there is a wide range of establishments in Addis Ababa offering accommodation from international hotel groups (Golden Tulip, Hilton, Marriott Executive, Radisson Blu and Sheraton) and locally operated hotels that operate large five-star hotels to small, locally run hotels that offer basic accommodation and services. Several new international and local brand hotels are also in the pipeline and are expected to be operational in the next few years.

Hotels of different star rates cater for different market segments. Business travellers usually target luxury (four and five-star) hotels (Yang, Wong, & Wang, 2012). Based on the grading scheme carried out in 2015 by MoCT in collaboration with the UNWTO, there are 79 star-rated hotels that fall within the 1 and 5 star categories in the city (MoCT, 2016). As shown in Figure 12.5, the majority of hotels in the city fall within the 1- and 3-star categories. Overall, the total numbers of hotel rooms and beds in the city are 5,343 and 6,518, respectively (MoCT, 2016). Nevertheless, while the hotel

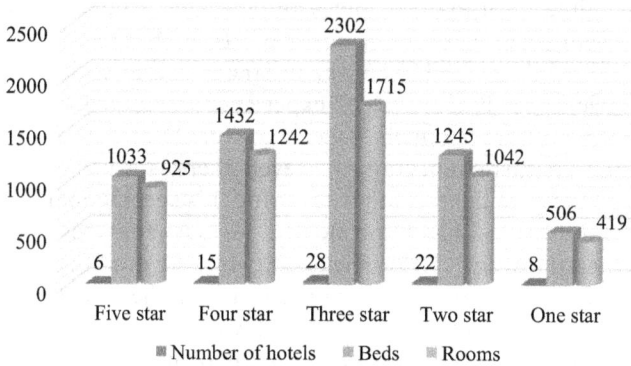

Figure 12.5 Number of star-rated hotels, beds and rooms in Addis Ababa, 2017.
Source: MoCT (2018).

industry is growing, the number of available hotel rooms is still the lowest by sub-Saharan standards. Regarding room availability, in 2017, Ethiopia is globally ranked 130 out of 136, compared to Uganda, Kenya and Tanzania at positions 43, 122 and 119, respectively (WEF, 2017).

Presence of Ethiopian airlines

Geographic proximity of delegates to the event-hosting destination is among the critical factors that influence success of convention sites (Crouch & Louviere, 2004). Equally important is airlines connectivity. In this regard, the presence of one of Africa's largest airlines, Ethiopian Airlines, established in 1946, and Addis Ababa Bole International Airport make the city an increasingly accessible and leading air hub in Africa (Otiso, Derudder, Bassens, Devriendt, & Witlox, 2011) with a high volume of transit passengers transiting to global and regional destinations. Ethiopian Airlines operates more than 20 domestic destinations and more than 90 international cities in Africa, the Middle East, Europe, Asia, North America and South America, and envisages reaching 120 international destinations worldwide by the year 2025 (Ethiopian Airlines, 2017). Being a major national and regional airline hub and gateway to Ethiopia, with good international connectivity and well-developed infrastructure, makes Addis Ababa a potentially decent location for an international convention and exhibition industry.

The rich attraction resources

Attractions are one of the prominent attributes of MICE tourism sector development (Swarbrooke & Horner, 2001; Tanford, Montgomery, & Nelson, 2012). Addis Ababa has abundant natural, historical and cultural resources such as Entoto botanical garden, historical buildings, churches, museums, diverse intangible cultural experiences and festivals. About 16 art galleries and 321 souvenir shops cater the visitors (ACTB, 2014). The presence of such diverse resources is essential in converting the MICE attendees into leisure spenders, enriching MICE attendees' experience and increasing their length of stay and expenditure (Connell, Page, & Meyer, 2015; Getz & Page, 2016b; Rogers, 2003). In this connection, however, there is a need for increased recreation and leisure facilities that cater for this specific market segment. The city administration has implemented various measures to utilize the existing potential (MoCT, 2015). The Entoto and surrounding tourist development project are intended to promote tourism in the city (Entoto Tourist Destination Development Office, 2017). However, despite such recent attempts, most of the events are not strategically planned to achieve tourism-based outcomes. This may be due to an exclusive focus on 'political' aspects of the meetings without making use of meetings as an avenue for sourcing visitors. Realising the full potential of the city and developing further tourism products would appear to be important means by which to enhance the city's MICE sector.

Challenges for the MICE sector in Addis Ababa

The MICE segment has encountered several challenges that include, but are not limited to, weak institutional and governance frameworks, lack of proper product development strategies, lack of trained human personnel, lack of research and development that informs tourism sector development, service quality and standards issues, lack of practice and learning platform, lack of proper database and insufficient support infrastructure, which are all discussed in the next section. These MICE sector challenges are rooted from the broader developmental problems in Ethiopia (and much of Africa for that matter), including high level of poverty, inadequate and poor-quality infrastructure, insufficient public facilities, poor sanitation and unemployment (UNHSP, 2017). The following section discusses these critical strategic challenges to MICE sector growth of the city.

Governance and institutional structure

Governance is the process of governing that involves the structures as well as the processes through which these structures exercise (Richards & Palmer, 2012). A destination that aspires to excel in event tourism needs to have a leadership that integrates stakeholders' interests and meets the broader goals of the destination without compromising the needs of individual events and their stakeholders (Richards & Palmer, 2012). Given that Addis Ababa is also the political capital of the country, both the national government and the city administration may have stakes in strategically directing the MICE tourism development and operation. There is no systematic integration of events with tourism components at times of planning and executing political events.

There is no specialized institution dedicated to MICE undertakings. With a growing number of meetings held in the city, and increased international trades and international relations, the current institutional system falls short in terms of providing dependable services. Instead, existing organizational structures, processes and procedures and existing relationships between different entities make it sometimes difficult to coordinate tourism and event activities in the city. Moreover, lack of trained personnel in the field and the absence of a formal event management-training program in the country contribute to the under-development of the sector. While coordinated collective actions are important to bridge the existing capacity, skill, infrastructural and investment-related problems, because of a lack of a tourism events strategy and absence of a convention office/bureau at the national, regional or city level that is responsible for leading and managing the MICE sector growth and development, the MICE sector is currently lagging.

MICE undertakings are married to tourism organizations without any dedicated departmental platform. Because of the lack of a dedicated convention office, for most events happening in the city, the Ministry of Foreign Affairs organizes an ad hoc committee that involves different sectorial offices. For example, this was the case during the 16th International Conference on AIDS and Sexually Transmitted Infections in Africa (ICASA) in 2011 at Addis Ababa, which was one of the largest meetings of its kind. The Ministry of Health was the owner of the event. Consequently, an ad hoc committee comprising Ministry of Foreign Affairs, Immigration, Security, Tourism, Ministry of Women's Affairs and Ministry of Youth and Sports was in charge of arranging the meeting. However, due to lack of one responsible organization, i.e. convention office, learning and practice sharing from such a meeting are not documented.

Moreover, due to a lack of clear procedure for conducting events, private event organizers are facing some challenges. This has been a common problem in granting permissions for domestic concerts and events. There is no clear procedure in terms of which organization gives permission, security and other city services for events. For example, few artists were denied permission to conduct music concerts because of bureaucratic procedures and security reasons (Addis Fortune, 2015). As a result, huge income opportunities of concerts that can also help in facilitating domestic tourism are not properly utilized. Similarly, provision of support to the private sector, ensuring implementation of basic standards (e.g. accommodation, transport) and conducting inspections of convention centres and events, has not been materialized. This requires a high level of synergetic commitment from the Government of Ethiopia, city administration, the private sector, international organizations and the community.

Infrastructure, facility development and service

The MICE sector is essentially a multifaceted industry, with its activities requiring, to a varying extent, many different players that encompass several single components of travel, trade, accommodation, transportation and other amenities and services (Mistilis & Dwyer, 2000). While several efforts are going on to make the city better and competitive to MICE undertakings, the number and quality of infrastructure facilities and services are still in the lowest state (Wubneh, 2013). The quality and availability of services and infrastructure facilities in the city are very low. The provision of public infrastructure has lagged far behind the growing demand (UNHSP, 2017). Despite the issue of reliability, significant facilities and services have reached the city, including mobile phone networks, rail link networks, electronic banking and credit cards (Worku, 2010), internet cafes, electricity and water, and asphalt roads, to mention a few. Despite such evidence of modernity, lack of electricity and power cuts in Addis Ababa can be a barrier to MICE sector development. Internal transport is also essential. And with 150 sedan taxis operational in the city (ACTB, 2014), despite recent initiatives aimed at standardizing services and fares, taxi services are dominantly traditional with no standard procedure (Wubneh, 2013). A reliable and quality taxi service that complements the MICE potential of Addis Ababa is yet to come. This involves professionalism and knowledge-based certification of taxi drivers (Hussein, 2016).

Standards and MICE sector performance

Developing and maintaining the service standards of the facilities and services are prevailing challenges in the MICE sector (Mistilis & Dwyer, 2000). In Ethiopia, MoCT has the overall responsibility for setting and maintaining tourism sector standards whereas the city administration is responsible for ensuring implementation of basic standard and conducting inspections in its respective region according to the criteria set by MoCT. However, there is no specifically designed grading and classification scheme for establishing and inspecting the event organizing institutions, MICE events and facilities in the country in general and in Addis Ababa more specifically. While rating of hotels is important for the MICE sector growth, the meeting halls and other conference facilities in the hotels are mainly inspected as part of the facilities and addressed within the general hotel grading scheme, without explicitly addressing the MICE sector requirements, i.e. specific criteria to rate the MICE facilities are not in place.

In addition to the standardization of core facilities and services, travel issues and policies pertaining to visa facilitation, protocol, customs and credit card facility are critical to the success and sustainable development of the MICE events in the city and beyond. In Ethiopia, visas are required for all visitors, except Kenyan and Djibouti nationals. Ethiopia launched an e-visa service for all international visitors to Ethiopia effective on June 1, 2018 (Ethiopian Airlines, 2018), which was ranked 46 out of 54 African countries regarding visa openness (African Development Bank, 2017). These services require coordinated efforts of different entities with specific focus on addressing the business tourism visitors' need.

Peace and security concerns

MICE sector growth is susceptible to international and domestic changes in the political, peace, safety and security conditions (Getz & Page, 2016a). Increasing political violence and terrorism threatens the growth of the MICE sector (De Lara & Har, 2008). The presence of the Somalia-based terrorist group al-Shabaab and their local terrorist group affiliates is one of the most significant regional stability and security challenges in the Horn of Africa in general and Ethiopia specifically. Ethiopia, being located in one of the most volatile regions of the world, is subject to the spillover effects of the social and political crises of its neighbours. Moreover, the country itself has been the victim of various political uprisings and crisis that have negatively affected the tourism industry. Historically political disorder in different parts of the country has influenced the tourism industry, and this is epitomised by the current (since 2015) Oromo and Amhara ethnic protest and the associated declaration of a state of emergency. This is an important factor, not only in providing a secure destination for hosting international political meetings and events but also in converting the political delegates into leisure tourists.

Image of the country

The MICE sector is becoming an increasingly competitive marketplace, where only the best-managed and marketed cities are likely to stabilize their growth. Despite the positive associations of Ethiopia with its rich history, authentic cultural and natural resources, and recent rapid economic development, the media continues to portray stigmatized images such as poverty, security concerns, desertification, famine and diseases (MoCT, 2009). There is a serious lack of more positive promotion within tourism information and marketing materials. This presents an unprecedented challenge to the tourism sector and the MICE sector specifically, and necessitates strategic marketing and advertising messages that balance the abundant resource of the nation with the existing dominant image problems. Assessments of the value of events in promoting a positive national image, tourism marketing in general, and co-branding along with other prime products and destination features are vital to harness the full potential of the event segment. MICE events could tell the world that Ethiopia is a safe place, and in order to do this, building positive relationships with the media is essential.

Lack of reliable and consistent data

Reliable tourism data are crucial to make informed and evidence-based strategic policy interventions (Dwyer, Deery, Jago, Spurr, & Fredline, 2007), and identify potential and emerging source markets and competitors. Currently, Ethiopia lacks reliable, comprehensive and timely tourism data, resulting in uniformed decision and misrepresentation in global rankings regarding its performance and potential (World Bank Group & MoCT, 2012). Specifically, there is limited evidence of the precise magnitude of the benefits and associated costs of MICE undertakings, thereby limiting the opportunities for informed policymaking and implementation (Dwyer et al., 2007). There is no properly developed database and statistical system, i.e. Tourism Satellite Account (TSA) to capture and monitor progress in the MICE sector. The contribution of events for domestic and international tourism growth, and its benefits are pronounced in different occasions without a proper data collection and database. A well-functioning national tourism statistics system with strong regional-destination coordination is essential to make an informed decision (MoCT, 2013). Moreover, none of the conferences and meetings held in Addis Ababa is strategically reviewed in terms of its strengths and challenges. Furthermore, data on delegate arrivals are often not collected, and there is a limited attempt to estimate the economic contribution of the MICE sector. Efforts to understand delegates' perceptions and satisfaction are not made. Development of a database of MICE events and activities involved is needed to monitor and guide its role towards sustainable development. This requires investigating the objectives from the demand-side (Getz, 2013), i.e. who attends these events and why, which is yet to be done in the Addis Ababa context.

Conclusion

By highlighting the key MICE tourism development trends, issues and challenges, this chapter has shed light on what the major developing-world cities like Addis Ababa could do to better tap and capitalise on their MICE tourism potential. Addis Ababa has immense MICE potential being a political centre for AU, UNECA and other international organizations, which is not yet adequately utilized. The city must build on the existing potentials by investing heavily on infrastructure and service development, creating and strengthening key MICE sector growth drivers and MICE specific institutions, incorporating strategic plans that strengthen institutional and planning capacity that deliver essential services to the residents and the business community. This requires the development of an effective event tourism strategy, a distinctive brand for the city that outlines the MICE potential, drawing on its resource endowments and improving the contribution of MICE to the life of its citizens. Moreover, developing MICE specific criteria and standards, and creating collaboration and coordination between the

MoCT and the city administration are essential for maintaining the standards and achieving sustainable benefits. In light of this, our analysis suggests that the creation of an institutional structure, i.e. a convention bureau that is committed to the management of the MICE sector in the city and beyond, is an important step forward. Capacity building and training on the MICE tourism concepts will be a crucial juncture of developing a vibrant institutional setting that guides the sector with the necessary knowledge and resources to effectively monitor standards in the industry and uphold service standards across the entire sector.

Notes

1 Based on 2015 currency exchange rates 1 USD was 21.2735 ETB.
2 Based on July 2018 currency exchange rates 1 Pound sterling was 36.48 ETB.
3 This ranking is based on a projection of a city's potential rate of change in 13 indicators across four dimensions: personal well-being, economics, innovation and governance.
4 It is computed through an examination of a city's performance based on 27 metrics using five dimensions: business activity, human capital, information exchange, cultural experience and political engagement.
5 Ethiopia was a member of the League of Nations and joined the United Nations on November 13, 1945.

References

Addis Ababa Culture and Tourism Bureau (ACTB). (2014). *Addis Ababa Culture and Tourism Growth and Transformation Plan (2008–2012 EC)*. Addis Ababa: Addis Ababa Culture and Tourism Bureau.

Addis-Africa International Convention and Exhibition Center. (2017). *Addis-Africa International Convention and Exhibition Center Officially Begins Construction*. Retrieved from http://www.aaicec.com/index.php/component/k2/item/19-addis-africa-international-convention-and-exhibition-center-officially-begins-construction.

Addis Fortune. (2015, Sep 29). Teddy Afro's Concert for Mesqel Cancelled. *Addis Fortune*.

African Development Bank. (2017). *Africa Visa Openness Report 2017*. Retrieved from https://au.int/sites/default/files/documents/32480-doc-visaoreport2017_web_12mai17.pdf.

Associations, U. o. I. (2017). International Meeting Statistics Report [Press release]. Retrieved from https://uia.org/sites/uia.org/files/misc_pdfs/pubs/UIA_stats_PR17.pdf.

Connell, J., Page, S. J., & Meyer, D. (2015). Visitor attractions and events: Responding to seasonality. *Tourism Management, 46*, 283–298.

Crouch, G. I., & Louviere, J. J. (2004). The determinants of convention site selection: A logistic choice model from experimental data. *Journal of Travel Research, 43*(2), 118–130.

Degarege, G. A., & Lovelock, B. (2019). Sustainable tourism development and food security in Ethiopia: Policy-making and planning. *Tourism Planning & Development, 16*(2), 142–160.

De Lara, R. A. A., & Har, C. O. S. (2008). Reassessing the need for the development of regional standards for the MICE Sector for the ASEAN and Asia Pacific Region. *Journal of Convention & Event Tourism*, *9*(3), 161–181.

Dwyer, L., Deery, M., Jago, L., Spurr, R., & Fredline, L. (2007). Adapting the tourism satellite account conceptual framework to measure the economic importance of the meetings industry. *Tourism Analysis*, *12*(4), 247–255.

Dwyer, L., & Forsyth, P. (1997). Impacts and benefits of MICE tourism: A framework for analysis. *Tourism Economics*, *3*(1), 21–38.

Elias, M. (Producer). (2016, 10/02/2018). Ethiopia: It's the 'political capital' of Africa, but how much of a melting pot of cultures is it? *Mail & Guardian Africa*. Retrieved from http://mgafrica.com/article/2016-02-11-ethiopia-an-african-melting -pot-of-cultures.

Entoto Tourist Destination Development Office. (2017). Entoto Tourist Destination. Retrieved from http://entototouristdestination.com/.

Ethiopian Airlines. (2017). Three New Flights to Three New Destinations in Just Three Days [Press release]. Retrieved from https://www.ethiopianairlines. com/corporate/media/media-relations/press-release/detail/795.

Ethiopian Airlines. (2018). Ethiopia Launches e-visa Service to All International Visitors [Press release]. Retrieved from https://www.ethiopianairlines. com/corporate/media/media-relations/press-release/detail/965.

Ethiopian Tourism Organization. (2017). Tourism: A National Priority. Retrieved from https://www.ethiopia.travel/about-eto.

Federal Democratic Republic of Ethiopia, Ministry of Culture and Tourism (MoCT). (2009). *Tourism Development Policy*. Addis Ababa, Ethiopia.

Federal Democratic Republic of Ethiopia, Ministry of Culture and Tourism (MoCT). (2012). F.D.R.E Vision, Mission and Values. Retrieved from http://www .moct.gov.et/index.php/en/2012-03-30-11-10-31/2012-04-05-09-50-24.

Federal Democratic Republic of Ethiopia, Ministry of Culture and Tourism (MoCT). (2015). *Addis Ababa Tourism Marketing Strategy (2016–2020)*. Addis Ababa, Ethiopia.

Federal Democratic Republic of Ethiopia, Ministry of Culture and Tourism (MoCT). (2016). *Tourism Statistics Bulletin (2012–2015)*. Addis Ababa, Ethiopia. Retrieved from www.moct.gov.et.

Getz, D. (2008). Event tourism: Definition, evolution, and research. *Tourism Management*, *29*(3), 403–428.

Getz, D. (2009). Policy for sustainable and responsible festivals and events: Institutionalization of a new paradigm. *Journal of Policy Research in Tourism, Leisure and Events*, *1*(1), 61–78.

Getz, D. (2013). *Event Tourism: Concepts, International Case Studies, and Research*. USA: Cognizant Communication Corporation.

Getz, D., O'Neill, M., & Carlsen, J. (2001). Service quality evaluation at events through service mapping. *Journal of Travel Research*, *39*(4), 380–390.

Getz, D., & Page, S. J. (2016a). *Event Studies: Theory, Research and Policy for Planned Events*. London: Routledge.

Getz, D., & Page, S. J. (2016b). Progress and prospects for event tourism research. *Tourism Management*, *52*, 593–631.

Hussein, A. A. (2016). *Service Quality Practices and Customer Satisfaction in Taxi Companies in Nairobi*. School of Business, University of Nairobi. Retrieved from http://erepository.uonbi.ac.ke/bitstream/handle/11295/99277/Service%20Quality

%20and%20Customer%20Satisfaction%20in%20Taxi%20Companies. pdf?sequence=1&isAllowed=y.

International Congress and Convention Association. (2017). *ICCA Statistics Report Country and City Ranking*. Retrieved from https://www.iccaworld.org/knowledge/benefit.cfm?benefitid=4036.

Johanson, D. C., & Taieb, M. (1976). Plio—Pleistocene hominid discoveries in Hadar, Ethiopia. *Nature, 260*, 293.

Jones, C., & Li, S. (2015). The economic importance of meetings and conferences: A satellite account approach. *Annals of Tourism Research, 52*, 117–133.

Kearney, A. (2016). *Global Cities 2016*. Retrieved from https://www.atkearney.com/documents/20152/436064/Global+Cities+2016.pdf.

Locke, M. (2010). A framework for conducting a situational analysis of the meetings, incentives, conventions, and exhibitions sector. *Journal of Convention & Event Tourism, 11*(3), 209–233.

McCabe, V. S. (2012). Developing and sustaining a quality workforce: Lessons from the convention and exhibition industry. *Journal of Convention & Event Tourism, 13*(2), 121–134.

Ministry of Finance and Economic Development (MOFED). (2006). *Ethiopia: Building Progress a Plan for Accelerated and Sustained Development to End Poverty (2005/2006–2009/10)*. Addis Ababa, Ethiopia.

Ministry of Finance and Economic Development (MOFED). (2014a). *Ethiopia: Review of Macroeconomic Developments (2002–2012)*. Addis Ababa, Ethiopia.

Ministry of Finance and Economic Development (MOFED). (2014b). *Growth and Transformation Plan Annual Progress Report for F.Y. 2012/13*. Addis Ababa, Ethiopia.

Mistilis, N., & Dwyer, L. (2000). *Information Technology and Service Standards in MICE Tourism*. Paper presented at the Journal of Convention & Exhibition Management.

Mitchell, J., & Coles, C. (2009). Enhancing private sector and community engagement in tourism services in Ethiopia. *Overseas Development Institute*. London, UK.

Oppermann, M., & Chon, K.-S. (1997). Convention participation decision-making process. *Annals of Tourism Research, 24*(1), 178–191.

Otiso, K. M., Derudder, B., Bassens, D., Devriendt, L., & Witlox, F. (2011). Airline connectivity as a measure of the globalization of African cities. *Applied Geography, 31*(2), 609–620.

Phillipson, D. W. (1998). *Ancient Ethiopia: Aksum, Its Antecedents and Successors*. British Museum Press.

Richards, G., & Palmer, R. (2012). *Eventful Cities*. London: Routledge.

Rogers, T. (2003). *Business Tourism Briefing: An Overview of the UK's Business Tourism Industry*. London: Business Visits & Events Partnership.

Rogerson, C. M. (2005). Conference and exhibition tourism in the developing world: The South African experience. *Urban forum, 16*(2–3), 176–195.

Rogerson, C. M. (2015). Unpacking business tourism mobilities in sub-Saharan Africa. *Current Issues in Tourism, 18*(1), 44–56.

Selassie, S. H. (1972). *Ancient and Medieval Ethiopian History to 1270*. Addis Ababa: United Printers.

Sisay, A. (2009). *Historical Development of Travel and Tourism in Ethiopia*. Addis Ababa.

Spiller, J. (2002). History of convention tourism. In K. Weber, & K. Chon (Eds.), *Convention Tourism: International Research and Industry Perspectives* (pp. 3e20). New York: Haworth.

Swarbrooke, J., & Horner, S. (2001). *Business Travel and Tourism*. London: Routledge.

Sylla, M., Chruściński, J., Drużyńska, P., Płóciennik, P., & Osak, W. (2015). Opportunities and limitations for MICE tourism development in Łódź. *Turyzm, 25*(2), 117–124.

Tanford, S., Montgomery, R., & Nelson, K. B. (2012). Factors that influence attendance, satisfaction, and loyalty for conventions. *Journal of Convention & Event Tourism, 13*(4), 290–318.

The World Economic Forum. (2009). *The Travel & Tourism Competitiveness Report 2009*. Geneva: WEF.

The World Economic Forum. (2017). *The Travel & Tourism Competitiveness Report 2017*. Geneva: WEF.

United Nations Human Settlements Programme. (2017). *The State of Addis Ababa 2017: The Addis Ababa We Want*. Nairobi, Kenya.

Weber, K., & Chon, K. S. (2002). *Trends and Key Issues for the Convention Industry in the Twenty-First Century*. Paper presented at the Convention tourism: International research and industry perspectives, New York.

Weber, K., & Roehl, W. S. (2001). *Service Quality Issues for Convention and Visitor Bureaus*. Paper presented at the Journal of Convention & Exhibition Management.

White, T. D., Suwa, G., & Asfaw, B. (1994). Australopithecus ramidus, a new species of early hominid from Aramis, Ethiopia. *Nature, 371*, 306.

Worku, G. (2010). Electronic-banking in Ethiopia-practices, opportunities and challenges. *Journal of Internet Banking and Commerce, 15*(2), 1.

World Bank. (2017). Ethiopia Overview. Retrieved from http://www.worldbank.org/en/country/ethiopia/overview.

World Bank Group, & Ministry of Culture and Tourism. (2012). *Ethiopia's Tourism Sector: Strategic Paths to Competitiveness and Job Creation*. (76576). Retrieved from http://documents.worldbank.org/curated/en/639061468257058697/Ethiopias-tourism-sector-strategic-paths-to-competitiveness-and-job-creation.

World Travel and Tourism Council. (2017). *Travel and Tourism Economic Impact 2017 Ethiopia*. Retrieved from https://www.wttc.org/-/media/files/reports/economic-impact-research/countries-2017/ethiopia2017.pdf.

Wubneh, M. (2013). Addis Ababa, Ethiopia–Africa's diplomatic capital. *Cities, 35*, 255–269.

Yang, Y., Wong, K. K., & Wang, T. (2012). How do hotels choose their location? Evidence from hotels in Beijing. *International Journal of Hospitality Management, 31*(3), 675–685.

Part IV

The future of events tourism

13 Reshaping metropolitan cities and creative tourism through artists' vision

Silvia Grandi and Chiara Bernasconi

Introduction

Cultural routes, itineraries, trails and walks are becoming more and more important policy instruments in creating cultural products in leisure, recreation and tourism opportunities and in renewing the image of a place. These can be the result of top-down processes and bottom-up ones where geographical knowledge of places, creativity and managerial capacities can be blended in reshaping the perception of places. These can be considered a subset of the broader category of events. In spatial term, their specificity is that they tend to develop in time and space along a line versus a point, but this simple spatial observation limits the social, behavioural and experiential potential of these activities. In general, the attention on these products has been widely taken into consideration by tourism studies in the latest decades either from a tourism management and planning policy perspective (Bianchini *et al.*, 1992; Grandi, 2005; Jafari, 2008; Smith and Onderwatwr, 2008, Evans, 2009; Pappalepore *et al.*, 2014) or from a rejuvenation of heritage and pilgrimage practice viewpoints (Selwyn, 1996; Schmude and Trono, 2003; Dallari *et al.*, 2009; Berti, 2013; Olsen and Trono, 2018). However, the focus of the investigation has been mainly on tourism rather than on the leisure, recreation and excursion activities that take place in the hometown or in a nearby location. Yet, Carr (2002) argues that tourism and leisure present commonality between the underlying influences that define how people behave during their leisure and tourism experiences. Yet, recreation and leisure activities occupy most of people's free time. These can express great potential when hybridised with the concept of creativity and art, and might help tourism and urban policies in balancing the needs of the local population and the needs of visitors, a critical element in managing the sustainable use of the city spaces. The traditional Martinotti (1993) model related to urban population differentiates in (a) *residents*, those that permanently live and work in the city; (b) *commuters*, those that travels to work over an appreciable distance, usually from the suburbs to the centre of a city; (c) *city users*, those that go regularly in a city, for instance, to work, to make use of services or to shop despite being non-residents; (d) *excursionist*, those that visit a city for leisure for less than 24 hours; and (e) tourists, those that visit a city for leisure, business or other reasons for more than 24 hours. This classification can be very useful to explain the relationship among people and the impact of culture, creativity, leisure and recreation activities in urban areas.

Being the city a central place, recalling the classical Christaller's model (Christaller, 1966), the behaviour, the numbers and the variety of people living in urban area create critical masses that can support a creative and cultural economy. Skilled consumers or creative consumers (Richards and Wilson, 2006), the creative class living in the city (Florida, 2012) as well as postmodern tourists and excursionists, city dwellers and other residents tend to seek actively their experience at the destinations they choose being this far or close by. Artists could be at the forefront in this process shaping material and immaterial urban atmosphere, developing new forms of art performances in different cities, continuing or challenging the tradition of avant-garde performances around the city and psychogeography walks (Grandi and Bernasconi, 2011). London Walks story shows the role of actors in creating a new way of visiting the city (Tucker, 2009); Berlin, Barcelona and Montreal show examples of other ways to create off-the-beaten-track areas and creative activities in urban areas (Maitland and Newman, 2014; Pappalepore *et al.*, 2014; Delisle, 2019).

Based on the concept of creativity, this chapter will explore the relationship between the sense of place, the role of itineraries as specific type of events and tourism, recreation and leisure, and art performances. In particular, the analysis will focus on the role of artists in creating itineraries and walks in metropolitan city that result in shaping the perceived sense of place of urban tourists as well as city visitors based on the use of the senses and ephemeral conditions rather than on historical facts and monuments. The case of Elastic City in New York City will be presented as an original way of representing the city as a changing organism, where the visitors are asked to actively participate in the creation of their experience.

Reshaping and innovating the sense of place in cities

In the last decades, territories have been facing an increasing global competition and effects result in "glocalised processes": on one side globalisation reduces distances – both in a physical and in an immaterial way – leading to homologation in merchandises, services and habits; on the other side this phenomenon triggers a renewed search for identities at a local level and a competition in diversification strategies (Dallari, 2006; Mariotti, 2012, 2015; Kong *et al.*, 2015). Developing or maintaining the uniqueness of a place can be achieved in a number of ways; events as well as thematisation, digitalisation and heritage mining are some of the most common strategies (Hall, 1989, 1992; Brown *et al.*, 2002; Carlsen and Taylor, 2003; Quinn, 2005; Dallari, 2006; Getz, 2008; Jafri, 2008; Avraham, 2014; Mariotti, 2015; Piva *et al.*, 2017). Themes can be used to link together a series of events as in the "festivalisation of the city" (Boogaarts, 1992). In order to hold audience attention, to attract visitors and to diversify the supply, festivals have to introduce new and innovative elements into their programmes. For this purpose they often compete directly with one another. Therefore, many major arts festivals have effectively become what we might call "creative spectacles". Creative spectacles are multifaceted in cause, ranging from supply factors, like cultural planning, tourism development and civic re-positioning, to demand factors such as serious leisure, socialization needs, lifestyle sampling and the desire for creative and authentic experiences. Performing arts and other festivals can now be considered a worldwide tourism phenomenon.

Cities should strive to reach their own distinctivenesses in order to continue to attract tourists and excursionists but, more than that, to satisfy residents, commuters and city users. Urbanisation processes are creating new needs for both residents and city users. Developing more than one distinctive aspect on the basis of intangible culture or creativity requires destinations to establish a link in the mind of the visitors and residents between particular manifestations of culture and creativity and specific locations. In addition, multisensoriality and personalization are recognized as fundamental elements of the postmodern society (Ferrari *et al.*, 2008).

Cities in particular need to develop creative spaces, spaces that are multifunctional, multitarget, dynamic, flexible and/or spaces of representations. Hall (2005) suggests that successful culture cities retain their creativity only by a constant self-renewing process thanks to people when they find a fertile milieu to do so. According to several authors (Hall, 2000; Florida, 2004; Evans, 2007; Maitland, 2007) artists, creative professionals, cosmopolitan consuming classes and migrant communities are the main drivers of this form of differentiation. The sense of space and place changes according to the eyes of visitors and their emotions (Urry, 1990, 2005) and it is well known that residents generally see and feel differently than city users or tourists. People, especially in this century, have a full range of needs, interests and cultural sensitivity that is extremely diversified. Evans (2007) argues that these cultural tours have little connection with creativity and contemporary places, and the sense of inauthenticity of the commodification of culture might outreach the true sense of places. Creativity and tourism in many aspects have conflicting interests similarly to residents and tourists in enjoying city spaces and services. Tourism tends to display an institutional culture that dislike diversity and distinction, risk and resistance, elements that are central in contemporary urban culture. Over planning in terms of branding and itineraries also tends to reduce the experience of the place and clearly becomes anti-creative (Evans, 2007).

On the other side, scholars that are more focused on endogenous growth factors and on local development perspective (Dallari, 2006) would argue that, for a long-term sustainable development of a city, heritage mining and slow territorial recomposition would be a necessary counterbalance to exogenous forces. Even if they recognize the value of creativity, they would claim the fundamental role of tradition and path dependency in identifying key factors of a sustainable evolution of a territory. Therefore, Dallari (2006) and Barone (2008) claim the central role of cultural itinerary as a space-time synthesis of historical heritage and novelty.

The innovative practices and creative solutions in today's leisure, recreation and tourism (LRT) represent a springboard towards the economic growth and well-being of the cities, regions and countries. Recalling Figini and Candela (2012) application of the classical love of variety model by Dixit and Stiglitz to tourism can be extended to all LRT stating that the tourist, as well as resident and city users, satisfaction increases with the variety of goods and services provided by a place. As a consequence, the demand for goods and services (accommodation, meals, entertainment, tickets for cultural events and visits) and the share of budget spent in the destination both increase with the richness of its variety. This implies that increasing the variety of goods and services, supporting the preservation and the availability of the diversity of natural environments, heritage and foster creativity activities focusing on creating experiences for visitor subjective and emotional states laden with symbolic meaning can enhance people happiness, sense of belonging and fulfilment. Therefore, tourism experiences should be projected as a way to reinforce a sense of personal identity or belonging to a community.

Conceptualising creativity for tourism, leisure and recreation activities

The word "recreation" is substantially and etymologically based on "creativity". Backing to Latin *re*: "again" and *creare*: "to create, bring forth, beget, generate", therefore this word highlights the sense of doing something creative or constructive, to cause to come into being, as something unique that would not naturally evolve or that is not made by ordinary processes; to evolve from one's own thought or imagination, as a work of art or an invention; to perform (a role) for the first time; to make by investing with new rank or by designating; to constitute; to appoint; to be the cause or occasion of; to give rise to; to cause to happen; to bring about; to arrange, as by intention or design.

In general terms, creativity represents the ability to imagine or invent something new, the ability to generate new ideas by using already existing information or to propose new ones by combining, changing or reapplying existing ideas and information (Grandi *et al.*, 2005). Creativity is also an attitude, i.e. to accept change and newness, a willingness to play with ideas, information and possibilities, a flexibility of outlook, the habit of enjoying the good, while looking for ways to improve it. Analyzing this concept it is easy to conclude that the more creativity is defined, the more creativity will be focused and limited. Instead, the more creativity is defined by focusing on how values, meanings, beliefs and symbols are shaped, the greater the chance that everyone's creativity will become less restricted and, therefore, able to release its potential. Each person can benefit from creativity, because it is a characteristic that belongs to human nature.

Given these elements, creativity can be considered the source which transforms creative ideas in new spaces and new experiences that are thought at the base of the creation of the kaleidoscopic sense of space and place (Grandi and Bernasconi, 2011). Therefore, creating is an imaginative, immaterial and an intellectual process as well as a physical, material one that can be recognized in all fields from traditional art fields (sculpture, painting) to invention, science, engineering, craftsmanship, cooking and so on. This is fundamental in all life-long people's lives, from childhood to silverhood, having beneficial health impacts, when experienced in a safe and healthy environment (UNDP, 2009).

Creativity and its results can be seen as part of an innovation process as the invention is strongly related to creativity knowledge and culture. However, to innovate is not just to invent. Recalling the Schumpeterian thoughts (Schumpeter, 1935), to innovate, in economic terms, means to do something new, useful and at the same time successful in reaching the market. "New" means something that does not exist yet, or new applications of something that already exists. This can be applied to places and spaces such as to metropolitan cities. The increasing metropolisation of the world population creates challenges and opportunities in recreation, tourism as well as urban planning. Congestion, commuting, environmental issues, multi-ethnic migration, degradation, empty industrial spaces, gentrification, creative classes and new technologies shape urban places and the way that people live in the city. Moreover, metropolitan cities are in strong competition among themselves to attract new firms, the creative class (Florida, 2002) and tourists. The consequence is that the importance of keeping the originality and uniqueness of the site to visit as well as to find strategies to ensure returns and loyalty. Nonetheless, finding occasion of recreation and leisure for residents and city users is vital in metropolitan urban centres to compensate urban negative externalities and promote quality of life. The ever pressing need to be innovative and competitive and the growing complexity and the need for changes in global markets are all conditions that call for creativity in order to not only find an answer to existing problems but also anticipate unexpressed demand of places, people and firms. Creativity and technology are two key assets of success for places, organizations and enterprises, both in local and global markets (Florida, 2002).

Within this perspective, creativity can be seen as a fundamental component in tourism product development as well as in recreation and leisure. In fact, creative activities can serve with very similar dynamics for tourists, as well as residents and city users. A creative activity can be part of a tourist experience that lasts much less than one day; it can last about one hour and it can be just a few blocks away from home, but expresses a novelty from the usual daily life, triggers emotions and uses skills that create a satisfying state of mind. Moreover, the evolution of creative tourism is moving from the development of small-scale creative experience and learning activities to the integration of tourism and creative economy (Duxbury and Richards, 2019). Creative LRT implies activities that offer people the opportunity to develop their creative potential through active participation in courses, learning experience, performative art actions that are characteristic of or are developed in the destination where they are undertaken. Therefore, creativity can be conceived as consisting of the following six elements:

- Creativity in the sense of art
- Creativity in the sense of craft (i.e. active making)
- Creativity in the sense of participating
- Creativity in the sense of generating
- Creativity in the sense of performing
- Creativity in the sense of reshaping the sense of places.

An important difference between creative and classic modes of cultural experiences generally consists of groups of visitors travelling with an expert guide who interprets the culture the tourist is seeing. The major difference between creative events, creative spaces and creative tourism is that these cases depend far more on the active involvement of tourists, and not just "being there", and passively listening and watching something. The development of creative experiences can also be considered a diversification strategy. Creative tourism as well as creative recreation and leisure events provides new ways of viewing a place and the self. This generates "experiential authenticity" for the participants, whether tourists or locals, as it serve as an occasion not only for learning something new but also for transformation of the self at its core, as one of the key ways to develop creative experiences, allowing participants to build their own narratives and to draw upon their own imaginative potential (Richard and Wilson, 2005, 2007, 2008a,b).

Itineraries, creative and performative urban walks and its evolution

Itineraries can be seen conceived as an event, as well as one of the basic units in developing creative tourism as well as creative recreation and leisure experiences. The concept of "itinerary" is linked to the notion of road, route, movement, a walk and it can be conceived as a planned route, a semi-organised path or a casual one. The basic is to connect two points: the origin and the destination that might be the same place in a circular trip. The choice of points and the path might follow a need, a theme, a discourse or an art performance, and this is what substantially characterises the creative process, the aim and the life of an itinerary. Yet, the consumption of an itinerary creates emotional states laden with symbolic meaning and these intense experiences can be a font of happiness, personal growth and self-realisation as well as reinforce the sense of belonging to a community. Planning an itinerary is the result of a complex decision process consisting of rational (for instance the shortest mathematical route) and irrational elements as an itinerary is a mental process based on the identification of features, either already existing or to be created, and which may result, for instance, in the case of cultural itinerary from both a historical analysis and a brand new design (Grandi, 2005; Barone, 2008). For a few years, the idea of "cultural itinerary" has been added to the wide range of "types" of cultural heritages, constituting the most complex one, both in its specification and identification and in its designing and handling. Moreover, a cultural itinerary is strictly connected to the landscape, which is an articulated cultural heritage containing further cultural heritages, creating a nested system of kaleidoscopicity based on heritage and other elements combined (Grandi, 2019).

Creative itineraries, instead, can be defined within a wider cultural itinerary concept where the expression of a creative activity by participants is key. For instance, active participation in a performative action – the walk – a creative itinerary or experiential route can be a creative tourist exercise . In postmodern society, in the experience economy (Pine and Gilmore, 1999) consumers seek strong emotional stimuli through subjective and customized consumption. This type of itinerary might be originated by an active participation in a performative action where senses and memories about a place play a central role in the itinerary's experience. The creative itineraries that are created by individual artists or small non-profit artistic organizations seem to hold a more authentic and real discovery of the city space.

Moreover, the creative potential of itineraries can be unearthed thanks to the irrational elements and artistic movements. As Coverley (2018) reviews, the idea of urban wandering relates to the older concept of the *flâneur*, theorized by Charles Baudelaire, some kind of Dadaism and Surrealism approaches on how to engage with the emotive force-fields that would permeate a city. Artistic movements such of the Letterist International and the Situationist International and other approaches such as Psychogeography, Fluxus and Soundscape have highlighted on how casualty, emotions and behaviour of individuals and itineraries are related. The basic idea is to create playful, sensorial and inventive strategies for exploring places, taking participants off their predictable tracks and trigger new awareness of the urban and rural landscape. Robert MacFarlane in *Psychogeography: A Beginner's Guide:*

> Unfold a street map... place a glass, rim down, anywhere on the map, and draw round its edge. Pick up the map, go out in the city, and walk the circle, keeping as close as you can to the curve. Record the experience as you go, in whatever medium you favour.
>
> (Coverley, 2018, p. 4)

In this approach rationality gives much more space to subjectivity and a chance to stress irrationality, emotions as well as the role of the senses. Thus rationality becomes not t only what can be seen by the eyes, but heard, touched, tasted and smelled. The component of the acoustic environment that can be perceived by humans has been highlighted in the Soundscape project. There is a wide history of the use of Soundscape depending on the discipline – ranging from urban design to wildlife ecology – but, in the creativity perspective explored in this chapter, the Soundscape suggests the importance of sensorial experiences in an immersive environment that can be revealed or (artificially) created for the leisure of people in closed spaces, urban spaces or rural ones.

The case study of Elastic City

Elastic City is emblematic in the analysis of the role and evolutions of creative walks, where the contribution of the artist in shaping the itinerary experience is especially important and is drawing from a particular use of the senses, a focus on personal memories, the use of ritualistic components and the observation of ephemeral elements in space that are normally not defining that specific space. From 2010 to 2016, Elastic City produced over 130 new sets of participatory walks, talks, ways and performances by over 100 artists. Todd Shalom, an America poet artist and the funder of Elastic City, created his first walk as a very personal walk, based on a visit back to the neighbourhood where he grew up. The experience was not only "nostalgic" but also meant to further explore how one can take his own personal experience and adapt it to something that he can give to others whom he may not know. A place that seems irrelevant to many is transformed in a point of interest thanks to the personal narrative that the artist reveals, thus proving that in a similar way, that each individual participant's stories matter. By legitimizing the artist's story, participants can relive their stories and understand the universality of their experiences in a similar way.

For example, during a soundwalk under the Brooklyn Bridge, participants are invited by the artist to get into pairs and help each other experience the world only guided by sounds. The sounds are not manufactured and controlled; they are the sounds of the environment that surround them and can be unpleasant, including traffic noise, the loud sound of trains on the elevated tracks, and sound effects generated by air conditioners in small alleys. The participant throughout the duration of the walk, by isolating sounds through exercises of walking blindfolded, increasingly comes to understand the richness and beauty of every sound around us. Moreover, the act of listening banal sounds that are usually blocked by the same participants during their daily lives functions as a meditation on their lives and mindful awareness of their surroundings. Sarah Owens, a rosarian and curator at the Brooklyn Botanic Garden, very interested in hedonistic pleasures and the sensuality of gardens, has led a group to walk through the rose garden linking the rose perfume to flavours and wines, broadening the participants' awareness of taste and smell in new ways, adding another layer of personal interpretation and complexity to standard rose garden tours.

Rituals play an important role in these creative walks. For instance, the artist puts together collective rituals for the group, or ask each individual to create one in the "4ever 21" walk; participants were led on a walk for eternal youth in Midtown, a neighbourhood in New York City mostly known for its endless array of options for shopping, eating and drinking. While many tourists stroll up and down Fifth Avenue for famous shops, few know what happens behind the scenes, in the beauty industry, and the walk was meant to make tourists and residents reflect on that, while making fun of our obsessions. The walk ended with the group making a necklace containing each participant's secrets for eternal youth, and placing it inside of a Forever 21 store, to symbolize the false commercial promise of eternal youth.

Because of the intrinsic nature of the walks taking place most of the time outside, in public spaces, there are always unforeseen events, like a cloudy sky on the day the shadows walk is scheduled, or presence of a certain sound on a sound walk. In these cases, the artist usually has developed plan B options, but also embraces the randomness and will ask the audience to do the same. Walks become and exercise in letting go of expectations and preconceptions about how a place looks and feels.

In most cases the element of chance is a welcome ingredient for the success of the experience. This draws from movements such as Fluxus, an avant-garde art movement that emerged in the late 1950s as a group of artists who had become disenchanted with the elitist attitude they perceived in the art world at the time. Fluxus art involved the viewer, relying on the element of chance to shape the ultimate outcome of the piece. Fluxus artists were most heavily influenced by the ideas of John Cage, who believed that one should embark on a piece without having a conception of the eventual end. It was the process of creating that was important, not the finished product. In a very similar way, the artist-led walks today need to be accepted by the participants as experiences that are unfinished and imperfect. They are at the end the ones drawing conclusions on what they went through. The artist often purposefully leaves them without a clear ending for their performative act.

Another important distinctive element of a creative artist walk is the political component. Most walks are developed and held in areas that are not very touristy and acknowledge changes in the areas that are controversial, without necessarily making it the focus of the walk. For example a shadow walk in Coney Island refers to the change of landscape not directly talking about the impact that developers and city actions have on the gentrification of the area – for example by pointing out the replacement of the original wooden boardwalk with cement and the action of turning the amusement park into one operator-led park, but by bringing attention to our internal and external shadows. Bringing attention to external factors starts first from an internal analysis of our own "masks" and ways in which we are not authentic with ourselves and with the people around us. The shadows walk by having participants innocently playing with the shadows created by lights blocked by objects around them suggests that they should think about how they are interacting with the world around them in a purposeful manner and building on the psychological interpretation of the shadow self. According to psychology, the "shadow" is the side of one's personality that contains all the parts of the self that people don't want to admit to having. It is at first an unconscious side.

The participant does not go on the walk to see art, but is part of the art itself (the performative act) and is shaping it with his/her action and participation. This is very apparent in the "Monumental walk", led by Niegel Smith, where the participants using their own experiences and the existing monuments as inspiration make new monuments to share with all who pass by. In this case the walk is undermining the concept of Monument. Who decides what is worth for everyone to "visit"? Why are certain structures erected to commemorate famous notable people or events at specific times? Most monuments that have been built by the powers at different points in history have been questioned in other times. The artist gives the power to the participants to respond to the so-called monuments, and see them with new critical eyes: do they speak to them? What kind of reaction do they instil in them? The unique challenge for the artist is to experiment in working with either an undiscovered place or a new perspective on a place, for example in the City Island Hop walk, looking at the City Island area in the Bronx, a strangely suburban area, where nobody goes if doesn't live there, or making a shadow walk in Coney Island, which is famous mainly for its amusement park.

Artist walks are also unique because the participants connect with the personal obsessions and interests of the leader, which is much more complex than the standardized knowledge of a tour guide, who has memorized a story to tell. Most of the times the tour guide's stories have been vetted before; they are somehow standardized and they don't contain particularly controversial information. They are mostly providing information considered important to know, or tell entertaining stories, but they are rarely critical. The artist-led walks instead are developed by professionals in their fields (graphic design, architecture, dance) who choose their passions above everything else, and their passion is a very tangible component in the way the walk is shaped in a personal way. Most of the time, they are not just providing information, but they are transferring insider knowledge and their passions. Connected to this, the focus of the walk is more on the development of one idea in depth, rather than touching on many areas (as the typical "touch and go" touristy experience), and the artist creates a really specific vocabulary of the walk. In the walk I led, I was able through the repetition and rehearsal of the walk to create a personal language to talk about shadows and encouraged the participants to create their own database of shadows through exercises. At the core of the development of the walk, there is experimentation, as opposed to proposing an already tested formula in the active walks.

During the creative walk a few factors that happen that make it an unforgettable/unique experience for the participants are connected to their ability to surrender to the artist leading it and to require active listening and participation – that's why talking about "participant" rather than "public" is preferred. The participant has to trust the artist as a guide. The artist becomes the expert in an apparently strange frame of knowledge that has forms of symbolic and spiritual knowledge. During these itineraries, the participant is asked to engage in activities that are normally considered un-important. His or her actions and response to the exercises that the artist creates make each participant become the artist in shaping the experience. Thus, the participant is invited to observe his actions and reflect on them, which is important especially in our fast paced environments where many of our daily actions become automatic.

Chance has an important role in the walk, as opposed to the active cul-tural walk where everything has been predetermined. The participants are asked to find connections and links with the reality around them. They are also asked to submit their own material and documentation, which goes to enrich the public communication of the project (social media play an in-creasingly important role in the creative walks, because they are promoting the participatory aspect). It is a two way documentation, open to get contri-butions for documentation from the public, in addition to the institutional images. The shift to create an experience that focuses on the visitor subjec-tive and emotional state is very clear in the artists-led walks. Tourists are more and more willingly becoming participants knowing they want to ex-perience their surroundings in a more authentic way. Accepting to be led by an artist on a tour of a city has many risks, including the one to be unhappy, sad, hurt, but never disappointed. It will never be a superficial or dull expe-rience; mostly it will question and potentially upset. What is the best way to really understand the characteristic of an area of New York City where many recent immigrants from Mexico live in Corona, Queens? Guided by women artists from Mexico, who only speak Spanish and who did cross the border without paper and are undocumented, the participants of the walk called "La Mano Inmigrante" (the immigrant hand) are presented with the unacknowledged truth that they are all immigrants. The group is then in-vited to imagine a neighbourhood where everyone's work and experiences are celebrated. The participants despite the state of world politics are able to make changes and a difference even when they normally feel powerless. They re-name subway stops, pass along their lives stories and perform the labour of people who, often underpaid and exploited, do the essential jobs that keep the city afloat. The walks become a political tool and have a much deeper impact in participants' lives than an entertaining informational tour on New York City.

Coherently with the ephemeral nature of the walks, Todd Shalom decided to end the organization that he had created with a final act. He organized a final three-week free festival throughout New York City featuring walks by seven artists/groups. "The Last Walk" was given seven times over two days in Prospect Park and featured cameo appearance by past walk art-ists. He decided to create a book called *Prompts*, detailing artists' prompts from Elastic City walks along with a guide to create your own participatory walks, as a sign that empowering each individual to create their own experience is the most important aspect of the process and goes beyond any other creative endeavour.

Conclusions

Cultural routes, itineraries and walks have been regarded as traditional products and policy instruments in creating cultural tourism products, as well as recreation and leisure opportunities in urban and rural areas. In the latest decades a significant push in renewing the traditional approaches developed , where creativity is conceived in the sense of art, craft (i.e. active making), participating, generating, performing and reshaping the sense of places. Designing or walking a route creates emotional states that artists can develop in to forms of art performances that blend into excursionist products consumed generally by a more complex set of people: city users, residents, excursionists, commuters and tourists.

The case study of Elastic City presented here reveals that participatory walks are working both at the level of the community and at the level of the self. Rather than a mass tourism product, walks can create "stress" relief and artistic projects. Moreover, the analysis of Elastic City over time shows that to keep the novelty and the authenticity of a creative urban walks implies a life cycle that tends to be limited in time and target a niche rather than a mass participation; thus it can be seen as a laboratory on post-modernity rather than as a fully successfully marketing innovation in eventing. However, the shift from a cultural to a creative approach highlights the importance of the relationship between the various people living a metropolitan area, the sustainability of tourist activities, the need of multisensoriality experiences and the role of the artists in shaping the perception of everyday spaces and thus shifting points of view.

Acknowledgements

The authors wish to thank Todd Shalom and the other artist of Elastic City for the interviews done. This chapter is a result of a common research: sections "Introduction", "Reshaping and innovating the sense of place in cities", "Conceptualising creativity for tourism, leisure and recreation activities" and "Itineraries, creative and performative urban walks and its evolution" have been written by Silvia Grandi, section "The case study of Elastic City" is edited by Chiara Bernasconi whilst the conclusions are the results of a shared work.

References

Avraham, E. (2014). Hosting events as a tool for restoring destination image. *International Journal of Event Management Research*, 8(1), pp. 61–76.

Barone, A. (2008). Le strade della cultura per lo sviluppo del Mediterraneo. In Sala, A.M., Grandi, S., and Dallari, F. (eds.), *Turismo e turismi tra politica e innovazione*, Patron, Bologna, Italy, pp. 175–184.

Berti, E. (2013). Cultural routes of the council of Europe: New paradigms for the territorial project and landscape. *AlmaTourism*, 4(7), pp. 1–12.

Bianchini, F., Dawson, J., and Evans, R. (1992). Flagship projects in urban regeneration. In Healy, P. et al. (eds.), *Rebuilding the City: Property-Led Urban Regeneration*. E&FN Spon, London, UK.

Boogaarts, I. (1992). La festivalomanie: A la recherche du public marchand. *Les Annales de la Recherche Urbaine*, 57–58, pp. 115–119.

Brown, G., Chalip, L., Jago, L., and Mules, T. (2002). The Sydney Olympics and brand Australia. In Morgan, N.J., Pritchard, A., and Pride, R. (eds.), *Destination Branding: Creating the Unique Destination Proposition*, 163–185. Butterworth-Heinemann, Oxford.

Carlsen, J., and Taylor, A. (2003). Mega-events and urban renewal: The case of the Manchester 2002 Commonwealth games. *Event Management*, 8, 15–22.

Carr, N. (2002). The tourism–leisure behavioural continuum. *Annals of Tourism Research*, 29(4), October 2002, pp. 972–986.

Christaller, W. (1966). *Central Places in Southern*. Prentice Hall, Germany.

Coverley, M. (2018). *Psychogeography (New Edition)*, Oldcastle Books, London.

Dallari, F. (2006). La competitività del turismo e lo sviluppo turistico locale. In Dallari, F., and Mariotti, A. (eds.), *Turismo tra sviluppo locale e cooperazione interregionale*. Patron, Bologna, Italy, pp. 65–74.

Dallari, F., Trono, A., and Zabbini, E. (2009). *I viaggi dell'anima. Società, Culture, Heritage e Turismo*. Patron, Bologna.

Delisle, M.-A. (2019). Montréal: A creative tourism destination? In Duxbury, N., and Richards, G. (eds.), *A Research Agenda for Creative Tourism*. Edward Elgar, Cheltenham, pp. 97–119.

Duxbury, N., and Richards, G. (2019). Towards a research agenda for creative tourism: Developments, diversity, and dynamics. In Duxbury, N., and Richards, G. (eds.), *A Research Agenda for Creative Tourism*. Edward Elgar, Cheltenham, pp. 1–14.

Evans, G. (2007). Creative spaces, tourism and the city. In Richard, G., and Wilson, J. (eds.), *Tourism, Creative and Development*. Routledge, London, pp. 57–72.

Evans, G. (2009). Creative cities, creative spaces and urban policy. *Urban Studies*, 46(5&6), pp. 1003–1040.

Ferrari, S., Adamo, G.E., and Veltri, A.R. (2008). Experiential and multisensory holidays as form of creative tourism. In Richards, G., and Wilson, J. (eds.) *From Cultural Tourism to Creative Tourism - Part 4: Changing Experiences – The Development of Creative Tourism*, Atlas, The Netherland, pp. 11–24.

Figini, P., and Candela, G. (2012). *The Economics of Tourism Destinations*. Springer, Berlin.

Florida, R. (2002). *The Rise of the Creative Class*. Basic Books, USA.

Florida, R.L. (2004). *Cities and the Creative Class*. Routledge, New York.

Getz, D. (2008). Event tourism: Definition, evolution, and research. *Tourism Management*, 29(3), pp. 403–428.

Grandi, S. (2005). Tourist itineraries and geographical information systems. Theory and applications. In Dallari, F., and Grandi, S. (eds.), *Economics and Geography of Tourism. The Opportunities Presented By Geographical Information Systems*. Patron, Bologna, Italy, pp. 137–148.

Grandi, S. (2019). La città caleidoscopica: gli itinerari urbani creativi dal turismo a forme di movimenti sociali. In Ferreira Cury, M.J., Magnani, E., and Cassia Pereira, R. (eds.), *Ambiente e território: abordagens e transformações sociais.* Editora Madrepérola, Londrina, Brazil, pp. 113–126.

Grandi, S., and Bernasconi, C. (2011). The kaleidoscopic sense of space and place: The eyes of the artists in route planning in metropolitan cities. In RGS Annual International Conference Abstracts, The Geographical Imagination, London, 30 August-1 September.

Grandi, S., Milazzo, V., and Callegati, E. (eds.). (2005). *Handbook on Innovative Thinking.* IPI, Roma.

Hall, C.M. (1989). The definition and analysis of hallmark tourist events. *GeoJournal,* 19(3), pp. 263–268.

Hall, C.M. (1992). *Hallmark Tourist Events: Impacts, Management, and Planning.* Belhaven Press, London.

Hall, P. (2000). Creative cities and economic development. *Urban Studies,* 37(4), pp. 639–649.

Jafari, J. (2008). Transforming culture into events: Faux pas and judicious deliberations. In *Proceedings of the International Tourism Conference Cultural and Event Tourism: Issues and Debates,* Ankara, Turkey, Detay Yayincilik, pp. 1–9.

Kong, L., Chia-Ho, C., and Tsu-Lung, C. (2015). Arts spaces, new urban landscapes and global cultural cities. In Kong, L., Chia-Ho, C., and Tsu-Lung, C. (eds.), *Arts Spaces, New Urban Landscapes and Global Cultural Cities: Creating New Urban Landscapes in Asia.* Edward Elgar Publishing, Cheltenham, UK, pp. 1–28.

Maitland, R., and Newman, P. (2014). *World Tourism Cities Developing Tourism off the Beaten Track.* Routledge, London.

Mariotti, A. (2012). Local systems, networks and international competitive-ness: From cultural heritage to cultural routes. Sistemi locali, reti e competitività internazionale: dai beni agli itinerari culturali. *AlmaTourism,* 1(5), pp. 81–95.

Mariotti, A. (2015). Città d'arte vecchie e nuove: le destinazioni del turismo culturale, In Salvati M., and Sciolla, L. (eds.), *L'Italia e le sue Regioni,* Istituto della Enciclopedia Italiana – Treccani, Roma, pp. 647–655.

Martinotti, G. (1993), *Metropoli. La nuova morfologia sociale della città,* Bologna, il Mulino.

Olsen, D.H., Trono A, and Fidgeon P.R. (2018) Pilgrimage trails and routes: the journey from the past to the present. In: Olsen DH and Trono A (eds) Religious Pilgrimage Routes and Trails: Sustainable Development and Management. Wallingford, CABI, pp. 1–13. DOI: 10.1079/9781786390271.0001.

Pappalepore, I., Maitland, R., and Smith, A. (2014). Prosuming creative urban areas. Evidence from East London, *Annals of Tourism Research,* 44, pp. 227–240.

Pine, B.J., and Gilmore, J.H. (1999). *The Experience Economy: Work is Theatre & Every Business a Stage.* Harvard Business School Press, Boston, MA.

Piva, E., Cerutti, S., Prats, L., and Raj, R. (2017). Enhancing brand image through events and cultural festivals: The perspective of the Stresa festival's visitors. *AlmaTourism,* 8(15), pp. 99–116

Quinn, B. (2005). Arts festivals and the city. *Urban Studies,* 42(5/6), pp. 927–943.

Richard, G., and Wilson, J. (2005). Developing creativity in tourist experiences: A solution to the serial reproduction of culture? *Tourism Management,* 27, pp. 1209–1223.

Richards, G., and Wilson, J. (2006). Developing creativity in tourist experiences: a solution to the social reproduction of culture. *Tourism Management*, 27(6), pp. 1209–1223.

Richard, G., and Wilson, J. (2007). *Tourism, Creativity and Development*. Routledge, London.

Richard, G., and Wilson, J. (2008a). The changing context of cultural tourism. In Richard, G., and Wilson, J. (eds.), *From Cultural Tourism to Creative Tourism*. Atlas, Arnhem, pp. 3–7

Richards, G., and Wilson, J. (2008b). Changing places, the spatial challenge of creativity. In Richards, G., and Wilson, J. (eds.), *From Cultural Tourism to Creative Tourism – Part 3: Changing Places, the Spatial Challenge of Creativity*, Atlas, The Netherland, pp. 7–10.

Schmude, J., and Trono, A. (2003). *Routes of Tourism and Culture. Some Examples for Creating Thematic Routes from Italy, Greece, Portugal and Germany*. Universität Regensbourg Wirtschaftgeographie und Tourismusforschung, Regensbourg.

Schumpeter, J. (1935). The analysis of Economic Change in *The Review of Economic Statistics*, reprint in *Readings in Business Cycle Theory*, 1944, Blakiston, Philadelphia.

Selwyn, T. (1996). Introduction. In Selwyn, T. (eds.), *The Tourist Image: Myth and Mythmaking in Tourism*. Chichester, London, pp. 1–32.

Smith, M., and Onderwatwr, L. (eds.). (2008). *Selling or Telling? Paradoxes in Tourism, Culture and Heritage*. Atlas Reflections, pp. 23–34.

Tucker, D. (2009). *London Walks*. Virgin Books, London.

UNDP. (2009). Origins of the Human Development Approach, http://hdr.undp.org/en/humandev/origins.

Urry, J. (1990). *The Tourist Gaze: Leisure and Travel in Contemporary Societies*. Sage, London.

Urry, J. (2005). The place of emotions within place. In Davidson, J., Bondi, L., and Smith, N. (eds.), *Emotional Geographies*. Ashgate Press, Burlington VT, pp. 77–85.

14 Counterculture and the future of music festivals and events

Daniel Wright

Introduction

> Countercultures are transgressive, avant-grade movements. The counterculture embrace of change and experimentation inevitably results in pushing beyond accepted views and aesthetics.
>
> (Goffman, 2005: 33)

What will music festivals and events offer to future market consumers? How will festival and event organisers meet the demands of consumers? The complexity of the modern human is deepening as society and the individual are receptive to a vast amount of ideas available to them. This can be attributed to growing global social interaction, and consequently, this is impacting culture. From a festival and events perspective this needs to be considered, as establishing markets and products based on new or transforming cultures is an opportunity for organisers. To explore this, it is necessary to consider the process of standardised culture and dominant forces such as globalization, Westernization, liquid modernity, commercial dominance and capitalism and a significant perspective taken in this chapter is counterculture, all as current drivers leading us into the future.

This chapter considers the future of the music festival and events industry, offering discussions surrounding the quest for uniqueness in a standardised commercial environment, taking a counterculture perspective. Gathering knowledge in which to identify counterculture movements and how these could be central to gaining a competitive advantage for future businesses are crucial. Unlike the past, society today has at its disposal an unparalleled amount of personal information, known as big data, and access to this large data source has the potential to be strategically decisive. To understand future festivals and events, one is required to explore wider social, cultural and political contexts. However, many of these now exist on social platforms. Big data on social platforms allow access to personal information, and those with access to it can begin to locate changes in society, locate patterns of change in cultural, social and political behaviour, attitudes, beliefs and more. These could be counterculture movements, moving parallel to mainstream interests which have the potential to become financial powerhouses, moving from the marginalised to mainstream consumerist society. Many counterculture movements have eventually become part of mainstream culture in some form. A significant factor that has restricted them in doing so quicker is time and social presence and exposure. In time, mainstream society has embraced counterculture movements such as the free speech movement, environmentalism, pop art, spirituality, gay liberation and various modern LGBT social movements. The counterculture of the 1960s often seen as an anti-establishment cultural phenomenon soon branched into different forms of expression, importantly, often surpassing the culture/counterculture divide (Gair, 2007). Countercultures have aimed to establish change to mainstream and well-formed systems in society. "Regardless of whether they are ultimately successful or doomed to failure, counterculture movements are a vital and a natural part of our existence" (Widewalls, 2016*). This chapter initially considers the importance of the music festival and events industry. It then explores culture, society and counterculture as an engine operating in the margins of society, before considering the impact of the internet and social media. Subsequently, the chapter discusses how big data analytics (BDA) offer an opportunity for festival and event organisers to recognise developing counterculture movements and how these have the potential to become profitable within the mainstream consumer environment.

The music festival and events industry

> We can see the fingerprints of a certain generation in the lyrics and sound of that time.
>
> (Huang, 2015*)

Music has been a significant pillar throughout human history. Today, popular music of our time is said to reflect our present culture. As noted by Peters (2016: 27) the role of music in society has been explored by geographers, sociologists, anthropologists and with cultural studies, where the lens has been on aspects such as music's ability to create a sense of place, the relationship between music and nation-state citizenship, the politics of consumption, the relation between ethnicity, identity, place and music and the global music industry. Music is a powerful tool, "ever since different cultures started to form, there has always been a place for rhythmic sounds that can communicate our feelings. In every age and civilization this particular form of expression has existed, and in so many varied styles" (Exploring Your Mind, 2018*). One way in which society engages with music is at festivals and events. Festivals and events have long been part of human society. They have evolved throughout time, originally sharing links to religious practices and beliefs, to culture, art and sports, and they exist on local and global levels. Motivations for implementing and running events have much coverage across the academic literature, from economical (profit-making), social (i.e. celebrating culture, history, religion) and political (such as the staging of mega events) perspectives (Van der Wagen and White, 2010). Richards and Palmer (2012) recognised the importance of events and how they are often perceived as improving the quality of life for destinations or people involved, providing economic and social benefits, increasing publicity and profile raising, generating business interest and new opportunities and networks.

As Getz's (2008) notes, events are an important motivator of tourism, and especially in recent years, developing and establishing events have become a prominent figure in the international tourism arena and consequently within the marketing plans for many destinations as a tool for establishing destination competitiveness. The growing focus of festivals and events as a fuel for tourism has potential for generating further touristic interest for destinations. The individuals participating in events have the "opportunity to assert their identities and to share rituals and celebrations with other people" (Raj, Walters and Rashid, 2013: 4). Festivals and events continue to flourish in many forms and the music industry has greatly benefited from this growth. The IFPI Global Music Report (2017: 3) notes that music is "being enjoyed by more people is more ways than ever before". From a UK perspective industry data suggest the core music industry in 2016 made an estimated economic contribution (Gross Value Added or GVA) of £4.4bn to the UK economy, supported by 142,208 jobs (CIC, 2017). According to Dean (2016) the number of festivals listed on festival website eFestivals jumped from 496 in 2007 to 1,070 in 2015. Furthermore, UK live music audience numbers in 2016 were 30.9m with 27m attending concerts and 3.9m going to music festivals (CIC, 2017).

The popularity of music festivals and events has clearly grown in the past few decades and various factors have contributed to this growth. From a supply perspective there is big money to be gained from organising music festivals. Similarly, festivals are more lucrative for music artists and bands than touring. From a demand perspective often driven by social media attention, music festivals are great value for money and the millennials are driving the market forward (Umbel, 2015). Music festivals and events in the future will continue to be a place where individuals can come together to share and express similar interests, values and beliefs. However, if destinations do continue to focus on music, then what type of music culture will society have to pull on? The interests of the youth generation could hold the key to what will determine future success in the festival and events industry. Thus, could a better understanding of today's counterculture movements be the answer?

Culture and society

According to Marxist philosophy, cultural hegemony is the domination of a (culturally diverse) society by a ruling class. The ruling class control and manipulate the culture of that society, its values, morals, beliefs, explanations and more. The intended aim is for the ruling classes world view to become the norm, one that is universally accepted, the dominant ideology within society. Consequently, culture is politically, socially and economically justified; the status quo is maintained. Social institutions play a significant role in maintaining the dominant ideological hegemony. In Gramsci's (1973) work and concept of hegemony, he identifies that the power of capitalism is to establish and maintain control. In the past, maintaining dominant cultures were enforced via violence and coercive intimidation. However, the concept of hegemony suggests that the bourgeoisie (middle class) maintains its position of power via its dominance over the language of culture (Gramsci, 1973). The effect is a worldview that becomes embedded and proliferates (Spracklen and Lamond, 2016) across societies and cultures. Consequently, the values of the middle class become the framework for all, and thus, capitalism continues to be incontestable and reproducible (Gramsci, 1973). As noted by Spracklen and Lamond (2016), the miserable nature of this is a continued cycle of oppression; in a social dominant hegemony, the oppressed are trapped in a social world in which their own actions serve only to prolong their own misery, so how does society challenge the social dominant hegemony?

In the *Ascent of Man* Bronowski (1981) notes that using imagination, reason, innovation and technology humans can change the world in unimaginable ways, and much of this is not down to biological evolution, but cultural change. Bronowski (1981) sees culture as the fundamental driver of change and as a consistent theme throughout time. Consequently, the human world is a place that can be programmed, fashioned and manipulated as a result of human activity, and as society develops, people are tasked with developing and making a new world, ideally one that is better to the one that precedes their existence. Bronowski (1981) observes culture as a means of broadening the human imagination, by transporting people's minds through time and space, allowing us to visualise ourselves in the past, present and into the future. According to Bronowski (1981) this cultural phenomenon, the ascent of man, has transpired over a few thousand years (more than one hundred times faster than biological evolution). Significantly, important debates remain amongst anthropologists with respect to how individuals and societies culturally change and evolve (Fennell, 2006). Cultural globalization often denotes a process in which the ideas, values and meanings of life are transmitted across time and space in a manner that different people withhold shared social practices. Such a process is often recognised as the mutual consumption of culture, as cultures gradually become diffused of variety and historical context, much to the impacts of popular culture, media and international travel.

Standardisation of culture

Modernity is driving standardisation. In a premodern world, societies lived and relied on the local rather than national and global support. Communication networks, people and communities consumed products and services that today would be deemed nonstandard. Products and services were not made to meet international specifications and legal requirements, and were not mass produced and rarely travelled beyond distant boarders. With the growth and development of the modern state and trans-border institutions, the standardisation of phenomena such as language, measurements, law, banking and finance took centre stage (Anderson, 1991 [1983]; Gellner, 1983). Likewise, food, art, sports and music became a product/experience for a global market to consume. This process of standardization is often attributed to the increased speed and impacts of globalization, where there is a tendency to focus on shared standards, seeking comparability and bridging norms, values, morals across worldly places that were once very much different (Barloewen, 2003; Meyer et al., 1992).

Consequently, consumer preferences, tastes and consumption patterns gradually became standardised nationally, and eventually globally. However, this is not suggesting that everything and everyone is and has become standardized, and such a movement continues to witness resistance. A useful example to consider here is Theodor Adorno's theory of mass culture. Adorno aimed to refute the notion that there is chaos and randomness in mass culture. Instead, what exists in mass culture is repetition and predictability, as the 'culture industry' becomes institutionalised. The aim is to establish a culture where people are unaware of their true existence as politics diminishes cultural imagination and creativity (Adorno, 1990; Laughey, 2007). Adorno focused on popular music to prove his theory. It is argued that popular music is a product of the standardizing effects of the culture industry (Storey, 2006). Whilst this might be the case for many pop-charts topping hits, it would be naive to rule out the great variety of music genres and festivals to meet the needs of these people. Significantly, events and festivals offer a place where people can share values and beliefs and the counterculture festivals of the 1960s were often recognised as an anti-government anti-political movement.

A consequence of standardisation is the resulting diminishing of cultural practices, knowledge, skills and craftsmanship, as the nonstandard is squeezed to the margins of social existence, and in some cases, it becomes obsolete (Eriksen, 2014). Claude Lévi-Strauss (1989 [1955]) framed his idea around life-worlds becoming obsolete by modernization. Such a modern onslaught would result in the diminishing of life's unique ways and worldly variety across societies and cultures. Standardisation will not result in the end of culture. However, traditionalist ways could be left in the past, but with the potential to be reborn in the future. At present, what could blossom is new, modern forms of culture, cultures that are more globally accepted.

But looming in the shadows will be alternative movements. Just like previous societies, there are often counter-interests, ones where individuals and collectives seek the alternative to that of the mainstream. Such marginalised cultural alternatives could even be past cultures and practices awaiting renewed interest, or they could be new cultural movements waiting to be born out of current and future consumer interests. These would be recognised as 'alternative to mainstream' and within these there could exist an array of values, beliefs and behaviours, ones that have the potential to be targeted and commodified. This continuous flux in culture is important; what is culturally dominant today may not be tomorrow and our cultural hegemony might not be embraced by tomorrow's society. The idea of 'liquid modernity' captures the notion of progressive cultural change. Bauman (2000: 82) provides this observation of our cultural world by suggesting,

> What was some time ago dubbed (erroneously) 'post-modernity' and what I've chosen to call, more to the point, 'liquid modernity', is the growing conviction that change is the only permanence, and uncertainty the only certainty. A hundred years ago 'to be modern' meant to chase 'the final state of perfection' - now it means an infinity of improvement, with no 'final state' in sight and none desired.

Bauman recognises that structure in society exists, but that change is permanent, that structures and dominant cultures evolve. Their change and evolution can be born out of counterculture movements.

Origins and influence of the counterculture movement

The counterculture movement (whose origins are attributed to the early mid-20th century) refers to subcultures whose values and norms of behaviour contrast significantly from those of mainstream society, and more so, often purposely operate and function in opposition to that of mainstream society (Saglam, 2014). In line with earlier discussions, which suggested that society was dominated by a hegemonically capitalistic language of culture, Gramsci (1973) argued that counterculture offered the potential for the working class to escape the hegemonically capitalistic dominance. In the 1960s there was a surge in anti-establishment movements across the United Kingdom and the United States gradually spreading throughout the Western world. The youth at the time opposed traditional values held by society and began to engage in non-violent protests. The emergence of television, cinema and news radio played a significant role in spreading the counterculture movement at the time. As a source of information and entertainment, media platforms fuelled the distribution of knowledge, ideas and news stories. This led to the growth of the counterculture movements, all of which played a significant role as a powerful tool for social influence and eventually led to cultural change, especially among the youth. In fact, the second half of the 20th century revealed the power of the media as the centre of most (especially mainstream) cultures, as advertising, television, video and music proved vital to society (Widewalls, 2016*).

> By utilizing all the benefits that come by controlling what people are exposed to on a daily basis, one is suddenly able to dictate a whole new set of norms and trends that define what passes as culture during that period of time. Since media was proved to be a weapon of all the people interested in dictating the current standards of culture, this also turned media into a perfect tool for those who desired to use it in different directions than the current ones.
>
> (Widewalls, 2016*)

Importantly, the counterculture movement ensured society confronted challenging subjects and matters. The counterculture movement of the 1960s had a significant role in shaping modern day society, leading to a less conservative approach. This resulted in attitudes (beliefs and behaviours) that are more liberal in some societies. Opponents argue that the movement was a leader in degrading traditional values, disregarding cultural values and resulted in an unruly violent society. The counterculture movement driven by youth was questioning and challenging the more traditionalist and historic beliefs and views held by mainstream society. Rojo and Harrington (2017*) suggest, "Counter-culture is essential to growth of culture, and while it can be shocking, disruptive, even painful at times, the wise know that the marginalized often lead the body politic toward a stronger equilibrium, a more perfect union". There is no doubt that the counterculture movement had a lasting effect on Western society, leaving its mark on the music industry, fashion and art, all of which have continued to be a key aspect of modern culture. Significantly, music festivals and events today are big businesses, and many of their origins can be traced back to the 1960s counterculture movements (Wolf, 2014). It is this lasting impression on society, the pressure to open society to new beliefs, behaviours and attitudes, that is significant. This lasting legacy of the counterculture movement is important because it opened society to new ways of being. As Goffman (2005: 28) suggests,

> Our defining vision asserts that the essence of counterculture as a perennial historical phenomenon is characterised by the affirmation of the individual's power to create his own life rather than accepting the dictates of surrounding social authorities and conventions, be they mainstream or subcultural.

The power struggle between the ability to create one's own life and one that is placed in front of them by society is central when considering a future events industry and the production and participation in future festivals and events.

A commercial cycle of culture: from counterculture to mainstream

> ...the economy is increasingly culturally inflected [while]... culture is more and more economically inflected.
>
> (Lash and Urry, 1994: 64)

According to Debord (1977) capitalist culture is a rigid game, where alternative, counter ideas are allowed in public discourse once they have become trivialized and sterilized of any potential corruptive influence on society. Thus, ideas counter to the mainstream are exploited and then reabsorbed back into the mainstream, "adding a pinch of spice to refresh older ideas" (Spracklen and Lamond, 2016: 16). Similarly, McGuigan's "cool capitalism - the neutralizing of dissent through the commodification of dissatisfaction and selling back to the dissenter as a depoliticized product" (Spracklen and Lamond, 2016: 16). Spracklen and Lamond (2016) suggest that the Pride-Festival fits neatly into the idea of counter to mainstream. Once a radical declaration of otherness, standing up against the dominant cultural forces, the festival is now a place of conspicuous consumption. Today, Pride-Festivals have become corporate, requiring ticketed entity, with large sponsors. Significantly, "the power of the pink pound constructs individuals not as social and political agents, but as economic actors who support a dominant hegemony that is still intrinsically LGBTQ-phobic" (Spracklen and Lamond, 2016: 16).

Whilst counterculture movements can impact on social-cultural beliefs, values, attitudes, with 'varying degrees of spice', their ideologies have the potential to be lucrative businesses. Locating the next popular music festival or event that could become mainstream can often be found in counterculture movements and their ideologies. As noted by Rojo and Harrington (2017*) "at any given moment a counter-culture is developing before your eyes. Authoritarian governments know this. So do, as it turns out, lifestyle brands, sociologists, and PR firms". In the 1960s ravers were part of the counterculture scene, people who could be described as lively, frequent partygoers. Raving, once an underground scene, taking place in unpopulated and uncontrolled environments, has now become a highly popularised and commercialised consumer experience. Anyone can participate in raves and entire music venues, and parks are purposefully transformed into "playgrounds for the electro-dance crazed partier" (Pérez, 2014*). The summer music festivals have become a place where people can escape from their routine lives, a place where they can explore their alternative selves. The hippie culture saw birth to the Isle of Wight Festival in 1970. Delistraty (2014) notes how Woodstock, a 'quintessential counterculture festival in upstate New York' is often recognised as the rise of hippie culture during the Baby Boomer generation and sparked a counterculture trend throughout the world. Modern festivalgoers continue to seek out the historical counterculture roots, with the Burning Man (a weeklong event in Nevada's Black

Rock Desert) being a modern example of a consumer festival where people can experience counterculture (Delistraty, 2014). Today's summer music festivals are frequently linked to the hippie practices and behaviours of the counterculture movements of the 1960s.

> The hipster is able to play the role of the bohemian while simultaneously wielding the resources of the bourgeois.
>
> (Delistraty, 2014*)

Sarna (2014) explores the Coachella Valley Music and Arts Festival counter-culture to mainstream phenomenon. Sarna notes,

> it's a melody of cultural phenomena, creating a connected world in which counter-culture is craved so much so, it's becoming mainstream. The subcultures of the 60s, 70s and 80s are still prevalent, but no longer elusive. There aren't so much new subcultures as there are evolved ones, taking style references in retrospective.

Also observing, "the rare and unique become commonplace once everyone has access to them" (Sarna, 2014*). As identified throughout, culture is a commodity to consume. The counterculture movement often driven by youth can lay the foundations for future culture, consequently future consumer experiences.

Figure 14.1 presents a framework exposing the stages in which counterculture movements can eventually become mainstream commercial products and services.

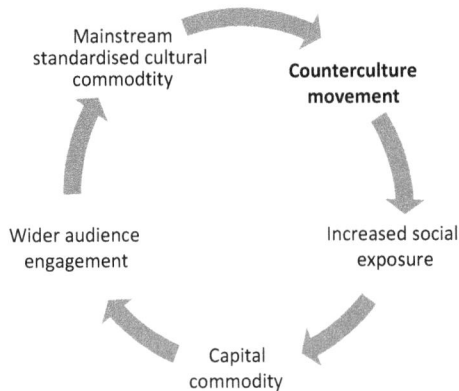

Figure 14.1 The cycle of commodified culture: from counterculture to mainstream culture.

Source: Author.

- *Counterculture movement* – a counterculture movement taking shape
- *Increased social exposure* – counterculture becomes exposed to wider audience
- *Capital commodity* – capital market exploits the movement
- *Wider audience engagement* – the wider public accept the counter movement
- *Mainstream standardised cultural commodity* – the movement becomes part of mainstream society as a product or service to consume.

Traditionally, as noted by McDowell (1997) there is an 'awkward relationship' between culture and the economy. As noted by Pérez (2014*)

the increasingly capitalized society we live in makes even a counterculture something entirely marketable. So, what happens when you commercialize individuality? Normally, it moves from the fringes and is subsumed into widespread culture; the digital age means that with an ever-increasing speed, counter-cultures have barely sprung up before they catch on and become mainstream.

As noted earlier, time was often a barrier to countercultures becoming mainstream; in today's world, with increased exposure through technology and social media, that transition could become much quicker for counter movements. Ultimately, culture, be it mainstream, popular, counter, traditional or other, can be a commodity – especially in more capitalist, consumerist societies. How this is packed and sold is key to attracting audiences.

The only way to support a revolution is to make your own

Future countercultures

...culture is useless unless it is constantly challenged by counter culture. People create culture; culture creates people. It is a two-way street. When people hide behind a culture, you know that's a dead culture.

(Yunus, 2010: 66)

With the introduction of the internet and the World Wide Web during the mid-1990s, talk of a social, cultural revolution began. The impact on politics, economics, culture and even the meaning of 'self' was all about to be transformed. The internet would have the ability to bring down organisations, decentralize control, globalize society and bring about a more harmonious, standardised way of living. As noted by Goffman (2005: 336)

before the popularization of the internet and the easy availability of other types of communication technology, most Western civilians didn't have convenient means for expressing themselves, or for distributing the results. Today, while free speech and thought, questioning authority, constant change, sexual liberty, and most other aspects of counterculturalness are not quite majority tastes, these liberties are permitted and available to a tremendous number of global citizens, if not the majority. Perhaps counterculture is no longer counter.

The internet would remove the traditions of 'men in suits' controlling and commanding the corridors of industry and business, and in their place, society would give rise to a new, modern 'digital generation', a self-sufficient, playful, colourful, accepting, a more whole individual, a generation of independent individuals built on a mixture of collaborative networks. The digital world would allow the individual to escape their physical appearance and establish themselves into new worlds of sociable connected spaces in which collectives could share communal interests (Negroponte, 1995).

   Pedersen and Mogensen (2011) explore the future of countercultures and suggest that present-day countercultures differ from the past. Whilst in the past, if one was part of a counterculture you were often fully integrated, it was your identity and it was often unlikely that you could also be part of another counterculture. They go on to note that current signals suggest that future countercultures will be even more complex. This is a result of entire worlds being connected via a pervasive (social-online) network of wireless internet and virtual worlds. Interestingly, and as a direct result of the online-social phenomena, the critical mass of individuals needed to make a counterculture viable and vibrant will not be limited to any particular physical location. Instead, it can be wide and spread all over the globe. Our world, one with increasing levels of change, at a faster pace, will result in new countercultures and subcultures consistently being born, or mutating, evolving from former movements and ideas. These will be further driven by the youth of the future, a generation of people who will want to try all types and sizes to see whether they could embrace them into their own personal identities or, increasingly so, just dip in and out of different countercultures driven by curiosity, novelty and to explore the diversity that will be available in future societies (Pedersen and Mogensen, 2011).

> Western and global culture today is a confusion of values. But who can deny that – within the chaotic complexity of this New World Disorder – even more individuals have increased individual freedom to nonconform to conventions and to communicate their own eccentric ideas?
> (Goffman, 2005: 336)

In the future, one will freely withdraw from a counterculture if it does not live up to expectations, unlike the past where one was deeply immersed, and withdrawing would have stronger social consequences and potentially a traumatic experience. The sheer quantities of complicated humans living on the planet will ensure that "cultural trends, counter (sub)cultures will multiply into an ungraspable myriad of forms, styles, and memetic aggregations" (Goffman, 2005: 336). The individual connection to any one culture will be looser than in the past and significantly the boundaries to enter and experience new cultures will be much more freely available. Significantly, some of these counter-movements will become mainstream and those who exploit this most effectively could be (festival and event) organisations who have access to big data.

Big data analytics: establishing future counterculture movements to gain a competitive advantage

> Land was the raw material of the agricultural age. Iron was the raw material of the industrial age. Data is the raw material of the information age.
>
> (Ross, 2016: 152)

BDA has grown as a popular topic of research among scholars and practitioners in recent years. Some scholars are going as far as to suggest that BDA is the "forth paradigm of science" (Strawn, 2012: 34), whilst Hagstrom (2012: 2) suggests it could be a "new paradigm of knowledge assets", and Manyika et al. (2011: 1) argue that BDA is "the next frontier for innovation, competition, and productivity". As discussed by Wamba et al. (2017: 1) these scholarly assertions are mainly driven by the "ubiquitous adoption and use of BDA-enabled tools, technologies and infrastructure including social media, mobile devices, automatic identification technologies enabling the internet of things, and cloud-enabled platforms for firms' operations to achieve and sustain competitive advantage". As Yiu (2012) notes, BDA ensures improved data-driven decision-making and innovative ways to organize, learn and innovate. Consequently, the aim is to strive for the strengthening of customer relationship management, improving the management of operation risk and enhancing operational efficiency and overall firm performance (Bean and Kiron, 2013, see Wamba et al., 2017). Hagel (2015) recognises how the business decision-making process is becoming reliant on BDA. Fosso et al. (2015) suggest that BDA should be a holistic approach for managing, processing and analyzing the 5 V data-related dimensions. The five dimensions are *volume, variety, velocity, veracity* and *value*. By assessing these 5 Vs, companies can create actionable ideas for providing sustained value, for measuring performance and importantly to establish a competitive advantage.

> Private companies now collect and sell as many as 75,000 individual data points about the average American consumer. And that is tiny compared with what's to come.
>
> (Ross, 2016: 153)

Big data is a term to explain how large amounts of information are used to understand, capture, analyse and even forecast trends in real time. Unlike the past where there was a lot of data captured, but limited systems in place for it to be cross-shared and analysed, the impressive value and functional use of big data now are its ability to make real time forecasts and efficient decisions. Consequently, companies, organisations, governments and people with access to such large data are able to foresee patterns and trends that otherwise might go undetected. With BDA, it will become increasingly possible for companies to identify and observe micro and macro (consumer) movements emerging and prevailing across the web, thus presenting correlations and patterns (between consumers) that were previously invisible (pre-web era) – and from a supply perspective, this is a game changer. The internet of things (IoT) can be best described as a network of physical devices such as mobile phones, computers, smart watches, heart monitors, cars, security cameras, smart meters, even fridges which are physical products fitted with sensors that are connected to the internet allowing the communication and sharing of information between devices – a digital sending and receiving of data. The number of connected products is increasing, with reports varying in estimation. A report by Gartner (2017) forecasted that 20.4 billion devices will be connected to the internet by 2020, whilst *The Guardian* (2018) estimates that by 2020, 30 billion devices worldwide will be connected to the internet.

Moving from the physical to the digital has become a common practice for the music industry in recent years. IFPI's Global Music Report (2017: 7) notes that:

> While physical sales remain significant in certain territories and for certain artists, there is no doubt that streaming is the key driver of growth, with the number of users of paid subscriptions having broken the 100 million mark and continuing to rise. Fans are engaged with music in an amazing variety of formats, from the vinyl revival to the phenomenon of musically, but the growth story is centred on services which are widening streaming's demographic appeal. Record companies and their distribution partners have been instrumental in this, licensing more than 40 million tracks to hundreds of digital services worldwide and developing the high-performance systems that allow music to be accessed around the world. Their approach has been global in scope and yet local in execution...

Online music streaming is creating large quantities of data. Future music festivals and event company's success could be enhanced with access to such data and providing experiences that meet consumer demand. Through BDA companies can identify counterculture (music) movements and significantly provide services and experiences to meet these emerging interests – purposefully influencing customer choice.

The potential future for the music festival and events industry: a counterculture approach

Throughout the past 50 years, society has seen a growth in the experience economy (moving beyond the manufacturing age) and within this consumeristic need for experiences, the festivals and events industry has established itself as an important element for destination stakeholders. The future will likely continue to benefit from individuals coming together, and music festivals and events will continue to provide a space where collectives can share their communal likes. Festivals and events today and in the future will maintain their ability to work as a "glue that bring people together through social cohesion, with joint plans and agreed strategic goals" (Derret, 2009: 109). Wolf (2014*) explores the relationship between todays music festival scene and their commercial aspects. She suggests that the summer music festivals and events "are discernable attempts to reunite pleasure with politics and retain the utopian ideologies that were sewn into the early Glastonbury, Isle of Wight and Woodstock festivals". The 1960s counterculture was anti-establishment and anti-corporate greed and anti-commercialisation. Today, UK music festivals like Leeds, Reading and T in the Park have arguably sold out to high-ticket prices, vast artist fees, sponsorship deals and commercial activity, whilst others (Boomtown, Shambala and Secret Garden Party) aim to retain counterculture origins. "The popularity of festivals like Secret Garden Party and Boomtown is not because of Burning Man, but because, like Burning Man, they understand the audience's desire to connect with the apparatus of production, rather than simply consuming products" (Wolf, 2014*). For example, Wales' Green Man Festival admits refusing sponsorship money from corporate companies and brands such as Wall's Ice Cream, suggesting that such a move is to favour local produce and to preserve the corporate marketing free atmosphere (Wolf, 2014). Wolf (2014*) identifies another important element about the counterculture music festival and event scene, that of participation. The popular lesser known "events encourage audiences to help in the creation of the event themselves, through building themed encampments, creating interactive art installations, and donning opulent costumes in accordance to annual, immersing themes". This is vital, as the counterculture movement was about identity, involvement and participation. Capturing this in today's world can be difficult, as commercial activity, productivity, growth and financial profit are deemed a vital ingredient for achievement. However, as shown by the success of events like the Burning Man, taking a counterculture approach, avoiding the lures of sponsorship and commercialisation, organisers can still realise success. However, for many, success will be realised if the transition from counter to mainstream is achieved, by applying the framework presented in Figure 14.1; the following is how big data could be applied to capture counter movements sooner and consequently establish a sellable festival or event.

- *Counterculture movement* – suppliers recognise a counter movement on social media by gathering information via big data.
- *Increased social exposure* – counter movement gains increased exposure on social media.
- *Capital commodity* – festival and event companies/organisers recognise the counter movement and begin to create and promote consumable experiences for the market.
- *Wider audience engagement* – the wider public buy into the counter movement as its popularity increases.
- *Mainstream standardised cultural commodity* – the counter movement becomes part of mainstream society and can be promoted and sold as a popular and standardized festival and event experience to consume.

Access to big data is the window into a world of counterculture movements.

Conclusions

This chapter explored the importance of counterculture from a music festival and event perspective. Globalization and social networks are establishing a world in which larger numbers of individuals can be part of a range of (alternative) cultures. Unlike the past, fully immersing oneself into a counterculture today is difficult. What is more likely to develop are opportunities for people to pick and choose cultural experiences. Consumers are increasingly likely to be presented with spectrum of culture experiences that range from mainstream to counter and they will be able to delve into them more freely than in the past. People today can select different lifestyles at the same time; we can pick and choose values and beliefs at different times of the year, when engaging with different experiences, even if they are contradictory to each other. We can attend different music festivals and events (which offer different social ideologies) with different family and friendship groups. BDA, the fourth paradigm of science', is moving society into a new era. With access to an immense web of information, organisers could create music festivals and events that mirror the shared values and interests of individuals. Future festival and event companies that have access to online big data could aim to use the information to identify counterculture movements and then bring to market appropriate experiences for individuals to share their interests (collectively). Goffman (2005: 30) suggests,

> our vision of counterculture centered on individuality is admittedly fraught with dangers. Many dissidents and counterculturalists have come to associate the word individualism with greed, selfishness, a lack of compassion, and the existential loneliness that comes from rejecting (or being rejected by) community. Countercultural individuality does not embrace mere self-centeredness. Counterculture individuality is deep individuality, shared.

Encouraged by the counterculture movement, Gramsci (1973) suggested that the oppressed in society should establish their own means of education, one that would encourage the appearance of class-connected intellectuals who would be able to articulate their own language of culture, a language that would provide a voice to the oppressed, eventually overthrowing the dominant hegemony. The power of counterculture lies within its potential to infiltrate the dominant culture in society, and consequently, allowing for culture change and for now liquid (late) modernity. The answers to our current challenges in society, like the past, could be eventually solved by the ideologies of counterculture movements that exist today. Thus, the power of music festival and events and the cultures that exist within them could be a powerful force for cultural change. Huang (2015*) suggests,

> culture and music flow together. What our parents used to dig, kids of today would deem as lame. And in a few years, the music we think is cool now will probably be outdated. It's nothing against the music. It's just a representation, a manifestation of what's constantly changing around us. With that said, we need to be very aware of our modern-day culture, but more importantly, we need to be intentional about the cultures we want to create and cultivate with our music.

The culture and identity of people and communities can change because of the music being produced and consumed. Therefore, it is important for organisers to understand that seeking counterculture music trends as a means of generating successful and profitable festivals or events could lead to wider social-cultural changes, especially if the counterculture ideologies are eventually not only absorbed but become part of mainstream culture.

Whilst this chapter focused on music festivals and events, additional research should consider the benefits of big data analysis for alternative festivals and events, for example foodism, ale festivals or even corporate events. Importantly, the use and application of BDA offer an opportunity to identify current trends and patterns in consumer choice and behaviour. For businesses, the ability of cultural and societal change suggests that movements and beliefs that where once deemed counter and peripheral can become the norm, the mainstream and accepted by wider society. It is often the case that in the margins of society the next big thing is brewing, waiting to take the world by storm. Those who recognise this have the potential of establishing (profitable) music festivals and event experiences based on counterculture movements.

References

Adorno, T. (1990) "On popular music", translated by George Simpson. In S. Frith and A. Goodwin (Eds.), *On Record: Rock, Pop, and the Written Word*, 301–14. London: Routledge.

Anderson, B. (1991 [1983]) *Imagined Communities: Reflections on the Origin and Spread of Nationalism.* London: Verso.

Barloewen, Von C. (2003) *Anthropologie de la modialisation.* Paris: Syrtes.

Bauman, Z. (2000) *Liquid Modernity.* Cambridge: Polity Press.

Bean, R. and Kiron, D. (2013) Organizational alignment is key to big data success. *MIT Sloan Management Review 54*(3), 1–6

Bronowski, J. (1981) *The Ascent of Man.* London: British Broadcasting Corporation.

CIC (2017) *Music Sector Size and Value.* Available at: http://www.thecreativeindustries.co.uk/industries/music/music-facts-and-figures/uk-music-market-size-and-value. Accessed on: 09.07.2018.

Dean, S. (2016) *Do the Growing Number of Music Festivals Actually Make Any Money?* Available at: https://www.telegraph.co.uk/business/2016/07/02/do-the-growing-number-of-music-festivals-actually-make-any-money/. Accessed on: 09.07.2018.

Debord, G. (1977) *Society of the Spectacle.* Detroit, MI: Black & Red.

Delistraty, C. C. (2014) *Commercializing the Counterculture: How the Summer Music Festival Went Mainstream.* Available at: https://psmag.com/social-justice/commercializing-counterculture-summer-music-festival-went-mainstream-86334. Accessed on: 25.02.2018.

Derret, R. (2009) How festivals nurture resilience in regional communities. In J. Ali-Knight, M. Robertson, A. Fyall and A. Ladkin (Eds.), *International Perspectives of Festivals and Events: Paradigms of Analysis*, 107–124. Oxford: Elsevier.

Eriksen, T. H. (2014) *Globalization*, 2nd Edition. London: Bloomsbury.

Exploring Your Mind (2018) *The Influence of Music on Our Lives.* Available at: https://exploringyourmind.com/influence-music-lives/. Accessed 30.05.2018.

Fennell, D. A. (2006) *Tourism Ethics.* Toronto: Channel View Publications.

Fosso, W. S., Akter, S., Edward,s A., Chopin, G., and Gnanzou, D. (2015) How 'big data' can make big impact: Findings from a systematic review and a longitudinal case study. *International Journal of Production Economics, 165*, 234–246.

Gair, C. (2007) *The American Counterculture, 1945–1975.* Edinburgh: Edinburgh University Press.

Gartner (2017) *Gartner Says 8.4 Billion Connected "Things" Will Be in Use in 2017, Up 31 Percent From 2016.* Available at: https://www.gartner.com/newsroom/id/3598917. Accessed on: 23.02.2018.

Gellner, E. (1983) *Nations and Nationalism.* Oxford: Blackwell.

Getz, D. (2008) Event tourism: Definition, evolution, and research. *Tourism Management, 29*, 403–428.

Goffman, K. (2005) *Counterculture through the Ages: From Abraham to Acid House.* New York: Random House.

Gramsci, A. (1973) *Selections from the Prison Notebooks.* London: Lawrence and Wishart.

Hagel, J. (2015) Bringing analytics to life. *Journal of Accountancy, 219*, 24–25

Hagstrom, M. (2012) High-performance analytics fuels innovation and inclusive growth: Use big data, hyperconnectivity and speed to intelligence to get true value in the digital economy *Journal of Advanced Analytics, 2*, 3–4.

Huang, B. (2015) *What Kind of Impact Does Our Music Really Make on Society?* Available at: https://www.thenewamerican.com/culture/item/17311-influential-beats-the-cultural-impact-of-music. Accessed on: 30.05.2018.

IFPI Global Music Report (2017) *Global Music Report 2017: Annual State of the Industry.* Available at: http://www.ifpi.org/downloads/GMR2017.pdf. Accessed on: 09.07.2018.

Lash, S. and Urry, J. (1994) *Economies of Signs and Space*. London: Sage.

Laughey, D. (2007) *Key Themes in Media Theory*. New York: Open University Press, McGraw-Hill.

Lévi-Strauss, C. (1989 [1955]) *Tristes Tropiques*. London: Picador.

Manyika, J., Chui, M., Brown, B., Bughin, J., Dobbs, R., Roxburgh, C., and Byers, A. H. (2010) *Big Data: The Next Frontier for Innovation, Competition and Productivity*. Seattle: McKinsey Global Institute.

McDowell, L. (1997) *Capital Culture: Gender at Work in the City*. Oxford: Basil Blackwell.

Meyer, J. W., Benavot, A., Cha, Y.-K., Kamens, D. H., and Wong, S.-Y. (1992) *School Knowledge for the Masses*. London: Falmer.

Negroponte, N. (1995) *Being Digital*. New York: Vintage.

Pedersen, K. K. and Mogensen K. Æ. (2011) *Countercultures of the Future*. Available at: http://www.scenariomagazine.com/countercultures-of-the-future/. Accessed on: 14.03.2018.

Pérez, M. (2014) *The 6 Most Commercialized Countercultures*. Available at: https://www.therichest.com/rich-list/world/the-6-most-commercialized-countercultures/. Accessed on: 05.07.2018.

Peters, P. (2016) Waltzing around the world: Musical mobilities and the aesthetics of adaption. In K. Hannam, M. Mostafanezhad and J. Rickly (Eds.), *Event Mobilities, Politics, Place and Performance*, 22–39. London, Routledge.

Raj, R., Walters, P., and Rashid, T. (2013) *Events Management Principles and Practice,* 2nd Edition. London: Sage.

Richards, G. and Palmer, R. (2012) *Eventful Cities: Cultural Management and Urban Revitalization*. London: Routledge.

Rojo, J. and Harrington, S. (2017) *When Counter-Culture Becomes Culture: Wastedland 2 and Andrew H. Shirley*. Available at: https://www.huffingtonpost.com/-jaime-rojo-steven-harrington/when-counterculture-becom_b_12646436.html. Accessed on: 25.02.2018.

Ross, A. (2016) *The Industries of the Future*. London: Simon & Schuster.

Saglam, B. G. (2014) Rocking London: Youth culture as commodity in the Buddha of Suburbia. *The Journal of Popular Culture, 47*(3), 554–570.

Sarna, K. (2014) *When Counter Culture Goes Mainstream*. Available at: http://theoriemag.com/when-counter-culture-goes-mainstream/. Accessed on: 25.02.2018.

Spracklen, K. and Lamond, I. R. (2016) *Critical Event Studies*, London: Routledge.

Storey, J. (2006) *Cultural Theory and Popular Culture: An Introduction*. Athens, Georgia: The University of Georgia Press.

Strawn, G. O. (2012) Scientific research: How many paradigms? *Educause Review, 47*(3), 26.

The Guardian (2018) *How the Internet of Things Can Grow Your Business*. Available at: https://www.theguardian.com/business-made-simple-with-vodafone/2018/feb/09/how-the-internet-of-things-can-grow-your-business. Accessed on: 23.02.2018.

Umbel (2015) *6 Factors Driving the Massive Growth of Music Festivals*. Available at: https://www.umbel.com/blog/entertainment/6-factors-driving-massive-growth-of-music-festivals/. Accessed on: 09.07.2018.

Van der Wagen, L. and White, L. (2010) *Events Management for Tourism, Culture, Business and Sporting Events*. Australia: Pearson.

Wambaa, S. F., Gunasekaran, A., Akter, S., Ren, S. J.-F., Dubey, R., and Childef, S. J. (2017) Big data analytics and firm performance: Effects of dynamic capabilities. *Journal of Business Research*, 70, 356–365.

Widewalls (2016) *Counterculture in Society and Art.* Available at: https://www.widewalls.ch/counterculture/. Accessed on 28.06.2018.

Wolf, J. (2014) *From 60s Counterculture to Big Business: The Politics of Festivals.* Available at: http://theconversation.com/from-60s-counterculture-to-big-business-the-politics-of-festivals-28682. Accessed on: 06.07.2018.

Yiu, C. (2012) *The Big Data Opportunity: Making Government Faster, Smarter and More Personal.* London: Policy exchange.

Yunus, M. (2010) *Building Social Business: The New Kind of Capitalism That Serves Humanity's Most Pressing Needs.* New York: Public Affairs.

15 Events tourism

A critical debate for the 21st century

Violet V. Cuffy and Bipithalal Balakrishnan Nair

Introduction

To date much of the debate in events tourism literature has mirrored developments in the field. Focus has largely been on leisure studies and related activities (cultural, arts or musical festivals; sports and recreational competitions; and private events) on one hand centred on issues of social and community involvement, satisfaction level, attendee's behaviour, identity and national cohesion (Draper et al., 2018). On the other hand, business events drew interest to matters of event planning, management, organisation, destination marketing and stakeholder concerns (Fenich & Fenich, 2016).

According to Getz (2008) in recent decades the events tourism sector has been the fastest-growing 'professional field' with tourists as a potential source market. Events, Getz suggests, are integral to national and cultural identity and contribute significantly to cross-cultural interactions, place authenticity, urban renewal, community building and regeneration that as demonstrated in this volume is closely linked to tourism development.

Similarly, Draper et al. (2018, p. 4) point out that "in the 21st century, global social media coverage of mega celebrations, festivals, sports competitions and vibrant private sector events pushed the event industry into centre stage, making events a staple of mainstream contemporary lifestyle". Arguably, satisfactory economic turnover from events boosts destination developments, favouring events tourism as a key national development strategy in many places.

This chapter turns attention to current trends that shape rapid shifts and innovative approaches for the future of events tourism in the 21st century. The impact of technological advances, big data, robotics, emerging economies, global political swings, climate change and the influence of a more knowledgeable customer can no longer be ignored. Increasingly key actors and policy stakeholders will need to embrace new frontiers such as breakthroughs experienced in virtual reality, augmentation, green events and authentic vacations, all now dictating how events tourism will be experienced in the future.

Following on Yeoman et al. (2013) forces of change that will lead the future of the events sector, we now explore ten key drivers identified for their

Table 15.1 Key factors affecting future events tourism

Key drivers	Implications for the future
The desire for authentic experiences	The drive towards more real and/or rural community-based events tourism and improved quality of service delivery without over packaging as a consequence of commodification
The future of luxury: wine food and music and the cultural capital experience	Addressing the needs of a more affluent and informed consumer with high quality taste and demands coupled with more disposable income facilitating a request for novelty and rich experience
The economic and political power of events	The inherent power of events to impact policy and national development, maximising its uses for socio-economic advancement of tourism
The environment and scarcity of resources	Impact of climate change on our basic and/or vital resources will need to be kept at centre stage when envisioning, planning and implementing events tourism at all levels across the globe
China's emerging middle classes	The rising China market as a significant player is shaping demand and product development across the globe
The Insperience experience and urban tribes	The importance of having up-to-date and state of the art technology and equipment within homes, well connected for hosting and entertaining of self and others
The future of prostitution	Hedonism and erotica have long been a pursuit of man, thus have always been a silent companion to the adventurous traveller and remain in demand.
Augmented reality and the information economy	Advances in information technology have led to shortening of the access gap in the sector with event goers opting for virtual experiences more and more; growth in this area is expected to be exponential
Ethic and generic engineering	Moral values are shifting quickly with generic engineering and artificial prototypes gaining more acceptance in the marketplace
I Robot	Robotics will continue to advance and invade our world for good or bad depending on your perspective; combined with artificial intelligence software new breakthroughs will continue to expand

Developed from Yeoman et al. (2013).

significance as current trends in the events world and by extension events tourism in (Table 15.1).

Scholars such as Getz (2008), Nelson (2009), Yeoman et al. (2013) and Pine (2014) among others examined matters of authenticity with increased focus on commercialisation and commodification. Concern with high rates of consumerism results due to contrast of focus with the modern-day traveller who seek real experiences for self-satisfaction and high levels of fulfilment rather than the experience of a product. Here, three main dimensions are envisioned: (i) the social anthropological dimension incorporating the cultural, regional or social identity of the events, (ii) community perception as a measure of local resident's involvement and participation as a matter of 'community control' and (iii) as a measure of tourist's perception. A critical present and future concern for the sector is how these together create authenticity in both the product and customer encounter.

The centre of economic and political power remains pivotal for large cities keen to host mega-events and the level of such power to impact the events tourism sector may be a game-changer for the succeeding decade. As demonstrated in this volume, events tourism developed around the Olympics for example can lead to significant city regeneration or become a national disaster if not planned, managed and delivered effectively. In an era of such global political shifts and shake ups among long standing established world unions, event organisers will need to be much more in tune with the impacts of political and lobbying trends. In addition, on a global level, geopolitical turbulence or cross-national political tensions are an undeniable consideration of this era and beyond. As such, Penfold (2019) who deemed Rio Olympics in Brazil as disastrous due to poor management advocates for complete alignment of economic and political power.

Events tourism continues to be designed, directed and underpinned by an era of technological revolution. Social connectivity, marketing trends, virtual worlds and key ICT developments in the area of transport and accommodation for example greatly influence organisation and management of the modern-day customer experience.

Furthermore, rapid and incessant improvement drives the industry because of great focus on the demand for constant capacity building and swift processing speed and task completion in all aspects of the sector. Event coordinators and advertisers are keen on knowing the intricate details of participants appearing on live feeds. With the alarming rise of widespread social media very little is private any longer. Access to information is instant at times pervasive. Moreover, continuous online innovation makes it incredibly easy for coordinators to accumulate information on fans, in order to tailor and enhance customer experiences.

In the world of big data, analytical skills become more essential for coping with the volume of information available about stakeholders and competitors alike. For example, the idea of purchasing tickets for others has been around for some time and introduces specific issues for those coordinators wishing to build knowledge database about fans. Notwithstanding, versatile ticketing is opening up new processes. Massive information can be assimilated and utilised for creation of much more focused promotional systems. Through investment and continuous product advancement marketers can now gather, process and dissect information in a less problematic way, making future advertising within a specific niche more efficient.

Science and technological innovations are not only shaping future events but also seen as a threat to current products and/or services. Virtual reality enables 360 degree perspectives. Accordingly, physical participation is often no longer essential. This has an array of implications for ease of access, costs, personal encounter and service experience. Individuals now often opt for virtual reviews of live events. Customers, Oreg et al. (2018) contend, now have the choice of attending their preferred events without the dread of passing up another significant opportunity. Furthermore, Yeoman et al. (2012) project that by 2050 virtual reality will overtake the real experience of watching in a stadium by providing better service encounters via Insperience.

Environmental concerns are now a worldwide problem. However, Zifkos (2015) and Laing and Frost (2010) lament the lack of much focused discourse on climate change in the sector. Regardless, Collins and Cooper (2016) have identified a significant shift in event research on environmental, social and economic impacts and sustainable issues. Nevertheless, this is quite a 'vexing issue' for the future of the events tourism sector irrespective of its type (festivals, sports, business) or size (small rural or mega).

Impact of consumer demand

Arguably, events tourism is increasingly one of the most important motivations for travel and plays a vital role in destination marketing, branding and development. Arguably, this could be as a direct result of the ever-expanding experience economy over the past 50 years (Yeoman et al. 2013). In that regard, Getz and Page (2016) put forward a set of consumer trends of relevance to this discussion for consideration in the area of events and festival as shown in Table 15.2.

Getz and Page (2016) build on the work of Yeoman et al. (2013) citing key consumer behaviours useful as a measure for the nature of future demand. An increasing appetite for the 'WOW' factor in events celebrations and festivals, increased affluence and choice of leisure activities, impact of the mobile era, active participation and authentic experiences, sustainability and the greening effect, nostalgia and cultural-social capital are all key phenomena observed and predicted as making continued significant contributions to customer experience.

From an industry perspective challenges of accessible tourism, random acts of violence by terrorists, shifts in weather patterns and sudden natural disaster of biblical proportions are a harsh reality looking towards the future. Inevitable matters of risk assessment and health and safety measures will come to the fore and remain central to events tourism going forward as policymakers and key actors grapple in dealing with these phenomena.

Moreover, megatrends such as the evolving visitor demand, sustainable tourism growth, enabling technologies and travel mobility bring on unforeseen challenges, threats and opportunities (OECD, 2018). As such the need for government policymakers, industry stakeholders as well as other actors to understand and address these within the local and international context will be crucial.

Table 15.2 Consumer trends shaping events tourism (in the area of events and festival)

Trend term	Summary
Everyday exceptional	An increase in celebration and the transformation of everyday experience into somewhat extraordinary and exceptional events
Magic nostalgia	A greater focus on reminiscence and celebration of the past events and festival
Leisure upgrade	The aspiration for leisure participation increases with affluence and events offer new form of social capital where participation is celebrated as an experience
Mobile living	We are living in more connected societies and living more connected lives which also transcends our leisure lives in which events (and non-leisure events) occur
Performative leisure	We are increasingly witnessing people celebrating their involvement in events and enjoyment through sharing the experiences via social media and mobile technology
Authentic experience	Consumers are seeking to accumulate more authentic leisure experiences and events and festivals offer one way to do this is increasingly through co-creation
Affluence	Consumers are becoming more demanding in terms of the needs and consumption within the experience economy
Ageless society	The rising age of the population in the developed world, due to greater life expectancy, has transformed the participation in events and festivals
Consuming with ethics	Consumers are starting to recognise the challenge of green issues and their own carbon footprint in everyday life and this may start to shape leisure consumption in the future around participation in events tourism
Accumulation of social capital	Consumers want to celebrate their achievement and participation in key events and festivals and this is part of the desire to accumulate experiences as part of this social capital repertoire

Source: Getz and Page (2016: 619).

On the positive side, the decade ahead looks exciting as innovation and entrepreneurship continue to add value to various dimensions of events tourism in practice. Advances in the digital revolution, augmented and virtual realities alongside the internet of things helping to create hybrid events; the demand for sophisticated venues such as green spaces and island venues; and the rise of festivalisation producing experiential events that are memorable and Instagrammable (CWT meetings and events, 2019) all allude to promising times ahead.

In conclusion, following on the work of Getz and Page (2016) arguably with the rise of events tourism as a subsector within the tourism industry a corresponding need of focused education and training for events tourism professionals and practitioners will become increasingly necessary as we look to the future in advancing the professionalism in the field and consumer experiences in practice.

References

Collins, A. and Cooper, C. (2016). Measuring and managing the environmental impact of festivals: The contribution of the ecological footprint. *Journal of Sustainable Tourism*, 25(1), pp. 148–162.

CWT Meetings and Events (2019). 2019 Meetings and Events Future Trends, Data Driven Insights and Expert Analysis to Maximise Your Results. Available at: www.cwt-meetings-events.com

Draper, J., Young Thomas, L. and Fenich, G. (2018). Event management research over the past 12 years: What are the current trends in research methods, data collection, data analysis procedures, and event types? *Journal of Convention & Event Tourism*, 19(1), pp. 3–24.

Fenich, G. and Fenich, G. (2016). *Meetings, Expositions, Events and Conventions.* Boston, MA: Pearson.

Getz, D. (2008). Event tourism: Definition, evolution, and research. *Tourism Management*, 29(3), pp. 403–428.

Getz, D. and Page, S. (2016). Progress and prospects for event tourism research. *Tourism Management*, 52, pp. 593–631.

Laing, J. and Frost, W. (2010). How green was my festival: Exploring challenges and opportunities associated with staging green events. *International Journal of Hospitality Management*, 29(2), pp. 261–267.

Nelson, K. (2009). Enhancing the attendee's experience through creative design of the event environment: Applying Goffman's dramaturgical perspective. *Journal of Convention & Event Tourism*, 10(2), pp. 120–133.

OECD (2018) Megatrends shaping the future of tourism, in *OECD Tourism Trends and Policies 2018*, chapter 2, Paris: OECD Publishing, pp. 61–91.

Oreg, S., Bartunek, J., Lee, G. and Do, B. (2018). An affect-based model of recipients' responses to organizational change events. *Academy of Management Review*, 43(1), pp. 65–86.

Penfold, T. (2019). National identity and sporting mega-events in Brazil. *Sport in Society*, 22(3), pp. 384–398.

Pine, J. (2014). *The Authentic Experience.* Montreal: Travel and Tourism Research Association.

Yeoman, I., Robertson, M. and Smith, K. (2013). A futurist's thoughts on consumer trends shaping future festivals and events. *International Journal of Event and Festival Management*, 4(3), pp. 249–260.

Zifkos, G. (2015). Sustainability everywhere: Problematising the "sustainable festival" phenomenon. *Tourism Planning & Development*, 12(1), pp. 6–19.

Index

Note: **Bold** page numbers refer to tables; *italic* page numbers refer to figures and page numbers followed by "n" denote endnotes.

Addis Ababa, Ethiopia: accommodation sector development 204–205; Addis Ababa Culture and Tourism Bureau (ACTB) 200; Addis Ababa Tourism Marketing Strategy (2016–2020) 201; Addis-Africa International Convention and Exhibition Center (AAICEC) 203–204; current state of convention and meeting centres 203–204; governance and institutional structure 206–207; image of country 209; infrastructure, facility development and service 208; international hotel groups 204; international tourist arrival 200–201, *201*; lack of reliable and consistent data 210; market segments 204–205, *205*; MICE tourism development 8; peace and security concerns 209; political role and presence of international organizations 203; presence of Ethiopian airlines 205; private sector involvement in event sector 204; rich attraction resources 205–206; sector potentials of 200–206; standards and MICE sector performance 208–209
adventure motorcycling, phenomenon of 105
adventure tourism 71, 105
Africa Bike Week 103, 107, **111**
African Union (AU) 196
Agricultural Development-led Industrialization (ADLI) programme 197
Albrecht, J.N. 7, 127

Amabile, T.M. 35
Ambelu, G. 8
America's Cup 185
Amore, A. 127
Andersson, T.D. 127
Anıl, O. 7
Annual Meeting of the Pacific Association of Paediatric Surgeons 137
Annual South African Toy Run, celebration of 106
ANOVA analyses 110
Anti-Gang Unit 76
Ark of Taste 90
Asia Oceania Society of Physical and Rehabilitation Medicine 137
Austin, D.M. 105, 106

Baby Boomer generation 239
Bacila, M. 172
Bakes, F. 5
Baksi, A.K. 62
Bang, H. 35
Barbieri, C. 175
Barget, E. 153
Barone, A. 220
Barrera-Fernandez, D. 8
Baudelaire, C. 223
BBB Rally 103
Bedate, A. 185
Bernasconi, C. 8
bidding for events: creation of long-term cultural strategies 185; cultural or undergoing restructuring 186; economy of cities or regions 184; high-rise events, celebration of 185; increasing

importance of 184–186; lack of long-term vision 187; negative impacts of 186–189; Porto 2001 187–188; and reactions in Plymouth, impact of 189–190; and social reactions 186–189; sociocultural change 188–189; tourism and marketing 187

big data analytics (BDA) 233; counterculture (music) movements 243; data-driven decision- making 243; "forth paradigm of science" 243, 246

Bikes, Blues & BBQ Festival 106

Biketoberfest 103

Blake, A. 153

BMW touring motorcyclists 106

Boley, B.B. 89, 91, 99

Booysens, I. 71

Borges, A. 15, 16

Botha, A. 70

Bowen, D. 55

Bozonelos, D. 168

Brida, J.G. 163

Bronowski, J.: *Ascent of Man* 234

Buning, R.J. 164

Burning Man 239–240, 245

business event tourism 3, 71; cultural events 164; definition 126; festivals and fairs 164; growth of 136; sporting events 164

Butler, R. 175

Candela, G. 220

Canterbury Earthquake Recovery Authority (CERA) 132

Canterbury Earthquake Recovery Commission 132

Canterbury Earthquake Response and Recovery Act 2010 131–132

Canterbury earthquake sequence 131

Cape Town: high rate of gangsterism in 76; introduction of Islam 73; Ramadan in Bo Kaap (*see* Ramadan)

capitalism 232, 235, 239

Carr, N. 217

Carter, E. 171

Cater, C.I. 106, 110, 118

CBD Recovery Plan 133

Central Business District (CBD) 72

CERTeT 149, 160

Chilembwe, J.M. 8

Chimpweya, J. 163, 173

Chi-square tests 114

Christchurch and Canterbury Tourism (CCT) 125

Christchurch Business District (CBD): quake effects in Christchurch 129–131; recovery of built environment in 132–134

Christchurch Central Development Unit (CCDU) 133

'Christchurch Central Recovery Plan' 133

Christchurch Conference Bureau (CCB) 130; first trade exhibition 135–136

Christchurch, New Zealand: after 2010/2011 destructive earthquakes 125; aftermath of Christchurch earthquakes (*see* quake effects in Christchurch CBD); CBD, recovery of built environment in 132–134; central city revitalisation strategy 134; conference industry 134–137; context and methodological approach 128–129; crisis-driven urban regeneration 134–137; earthquake aftermath 125; economic feasibility 133; natural disasters 126; post-disaster urban regeneration process 129, 138; public funding 136; quake effects in Christchurch CBD 129–131; quake-shattered Christchurch 134; recovery and rebuild 131–134; recovery governance and legislation 131–132; recovery process and multiplier effects 137; Red Zone and Green Zone 133; 'Re:START' mall 125; rise of conference industry in global cities 126–128; slow-burning stresses 126–127; strategy of telling positive stories 135; tourism recovery 134–137; transformation of conference industry in 7

Christmas Lights Switch On events, or book signings 22

Cicero, L. 7

cities (metropolitan) and creative tourism: Christaller's model 218; city users 217; commuters 217; "creative spectacles" 219; creativity for tourism, leisure and recreation activities 220–222; Elastic City, case study of 224–228; excursionist 217–218; itineraries, creative and performative urban walks and evolution 222–224; reshaping and innovating sense of place 218–220; residents 217; skilled consumers or creative consumers 217; through artists' vision 217–228

city rebranding *see* bidding for events
Civil Defence Emergency Management
  Act 2002 131
Club Cruisers 107
Coachella Valley Music and Arts
  Festival 240
co-creators 27, 63, 64, 165
co-entrepreneurship 96, 101
Coetzee, W.J.L. 5, 7, 127
Coles, C. 200
Collins, A. 164, 254
Collins-Kreiner, N. 174
commercial dominance 232
community-based events: citizen control
  48–49; meetup 49; role of users 49
Community-Based Tourism (CBT) 78;
  characteristics of 80; education and
  capacity building 80; elements for **80**;
  goals of sustainable development 79;
  at Komjekejeke 7; Quality-of-Life 79;
  success of 81
community control 253
community events 164
Cook, J. 189
Cooper, C. 254
Corbau, C. 171
corporate events 169, 247
countercultural individuality 246
counterculture movement: anti-
  establishment cultural phenomenon
  233; attitudes (beliefs and behaviours)
  238; big data analytics 243–244; capital
  commodity 246; commercial cycle of
  culture 239–241; commodified culture,
  cycle of *240*, 240–241; culture and
  society 235; future countercultures
  241–244; to gain competitive
  advantage 243–244; increased social
  exposure 246; LGBT social movements
  233; mainstream standardised cultural
  commodity 246; music festival and
  events industry 233–234, 245–246;
  origins and influence 237–238;
  personal information 232–233;
  standardisation of culture 236–237;
  television, cinema and news radio,
  emergence of 237–238; wider audience
  engagement 246
Coverley, M. 223
creative tourism 222
creativity: activities 221–222; and
  art, concept of 217; creation of
  kaleidoscopic sense of space 221;
  definition 220–221; elements 222;

itineraries 223; knowledge and
  culture 221; and technology 221; and
  tourism 219
crisis-resistant tourists 75–76
critical event studies (CES) 48
Crompton, J.L. 108
Cuffy, V.V. 5, 9
cultural events 164, 169
cultural globalization 234
cultural itinerary 220, 223
cultural tourism 71
culture and economy 241

Dadaism approach 223
Dallari, F. 220
Dangor, S. 73
Daries-Ramon, N. 6
Darwin, C. 189
David, N.M. 75
Davis, L. 16
Daytona races 103
Dean, S. 234
Debord, G. 239
Deery, M. 40
Del Barrio, M.J. 185
Delistraty, C.C. 239
Department of Arts and Culture
  (2017) 78
Department of Sports and Recreation 71
destination(s): attractiveness 20;
  branding 196; and business image
  formation 47; retail centres as 18–20
destination Israel and Malawi
  wanderlusts: Christmas markets 164;
  critical characteristics of niche events
  167; critical issues in tourism event
  destination 163–171; destination
  branding 167–168; destination
  competitiveness 165–166; Destinations
  Management Organisation (DMOs)
  plan 167; environmental externalities
  164; event tourism and travellers
  to Israel 171–172; festivals 169;
  investment decision-making 167; issue
  of festivals and events 166; leveraging
  strategies 166; niche tourism 175–176;
  sustainability of festivals and events
  167; wanderlusts and religious tourism
  events 172–175
destination management organisations
  (DMOs) 47–48
Destination Shopping Centres
  (DSC) 13; artificial mini-cities
  13; event portfolios as marketing

strategy 24–27; Meadowhall in
Sheffield or Westfield Stratford in
London 21; at Meadowhall, Trafford
Centre and Centre: MK 13; mini cities
5; proposed event portfolio model
28–29; using non-traditional events 13
Destinations Management Organisation
(DMOs) plan 167
DeWalt, B.R. 55
DeWalt, K.M. 55
digital generation 242
digitalisation 218
Dolnicar, S. 75
Domingues, A. 188
Doods, R. 79, 80
Drake, F. 189
Draper, J. 251
Dredge, D. 100, 118
Dubai Shopping Festival 13
Dutch colonialism 73
Dynamic Interaction Leisure 105
Dziewanski, D. 72, 73, 76

Earth Market, Şile 7; decision-making
and implementation processes 86;
food-themed events 86; slow food
movement and 90
earthquake aftermath 7, 125
eco-tourism 71
Edinburgh Festivals 39, 146, 154
Elastic City, case study of: active
listening and participation 226–227;
chance 227; creative artist walk 225,
226, 227; Fluxus art 225; intrinsic
nature of walks 225; "Monumental
walk" 226; past walk artists 227;
rituals 224–225; role and evolutions
of creative walks 224; shadows walk
225–226; during soundwalk 224; stress
relief and artistic projects 228
electronic word-of-mouth
(eWOM) 49, 62
Elefantentreffen (Germany) 103
empowerment: economic empowerment
89; framework 7; political
empowerment 89–90; psychological
empowerment 89, 94–96; Resident
Empowerment through Tourism Scale
89; social empowerment 89
entrepreneurship development:
analysis and construction of
framework 92; data collection and
survey 91–92; empowerment 94–96;
empowerment for sustainable tourism

development 89; event tourism and
sustainable tourism development
87–88; findings 92, 94; impact of food-
based event 7; research design 91;
rural entrepreneurship 88–89; sample
92; slow food movement and the
Earth Market 90; study setting 90–91;
supporting rural entrepreneurship
(*see* rural entrepreneurship)
environmental sustainability 170
Ethiopia: Addis Ababa, challenges and
sector potentials (*see* Addis Ababa,
Ethiopia); currency exchange rates
211n2; development of tourism
sector 196–197; examination of
city's performance 211n4; GDP
198; International tourist arrivals,
2000–2016 198, 199, *199*; international
visitors for conferences and
business purposes 200, *200*; MICE
tourism development (*see* MICE
(Meetings, Incentives, Conferences
and Exhibitions) sector); National
Tourism Development Policy (NTDP)
197–198; tourism sector development
in 197–200; Travel and Tourism
Competitiveness Index (TTCI)
ranking 198
Ethiopian Tourism Organization
(ETO) 198
European Bike Week (Austria) 103
European Capital of Culture
(ECOC) 184, 185, 187, 193;
designation of 185
European Commission 187
Evans, G. 219
event activity: fast-paced dimension 36,
41; short-term organisations 36, 37;
temporary organisations 36–37
event marketing researchers 118
event organisation 36; major events,
definition 36; pulsating nature 36
event populations 126
event portfolios as marketing strategy:
Alton Towers or Flamingo Land,
Retail Centres 25; city centres and
destinations 25; community-driven
events 26–27; Kim Priest, Marketing
Manager At Centre: Mk In Milton
Keynes 27–28; Marketing Managers
of Retail Centres 24, 26, 27;
Marketing Mix 25; Trafford Centre
or Meadowhall 24, 26; Westfield and
Bluewater 25

event teams: career-and relations-
oriented heterogeneity 38, 43;
casual employees 39–40, 42; full-
time and permanent employees 39,
41; functional heterogeneity 38,
39; heterogeneity 37, **38**; unpaid
volunteers 40, 42; Visitor Information
Centres 39
event tourism 71; complexity of 4;
contacts (followers and followed)
49–50; critical event studies 4–5;
economic and political power
253; environmental concerns 254;
event coordinators and advertisers
253; eWOM 49; factors affecting
future events tourism **252**; future
of 8; impact of consumer demand
254–255; investment and continuous
product advancement marketers
253; megatrends 254, **255**; rise
of festivalisation 255; science
and technological innovations
253–254; social capital 4; socio-
cultural and environmental impacts
of events 3–4; theorisation 4;
2050 virtual reality 254; 'WOW'
factor 254
event workforce management: event
activity 36–37; event organisation
36; event teams 37–41; inward
perspective 33–34; job satisfaction and
experiences 34–35; organisational and
staff perspectives 33–34; outward-
looking 33
eWOM *see* electronic word-of-mouth
(eWOM)
Expanded Public Works Programme
(EPWP) 78
Expo 2015 survey: Expo Milano 2015
143, 146; hypothetical increase in
consumption *150*, 150–153, **151**,
152, *153*; impact for Italy 148–157;
regional impact and inter-regional
redistribution of wealth 153–157;
survey 146–147; visitors spending **148**,
148–149
Expo 2015 visitors' expenditures, Italy 7–8

Facebook 51
Faisal, A. 7, 127
Ferreira, S. 71
festivals 3, 169; Bikes, Blues & BBQ
Festival 106; Coachella Valley
Music and Arts Festival 240; Dubai
Shopping Festival 13; Edinburgh
Festivals 39, 146, 154; and fairs 164;
Green Man Festival 245; Isle of Wight
Festival 239; Komjekejeke traditional
festival 76–81; music festival and
events industry 233–234, 245–246;
Neighborhood Theatre Festival 192;
Pride-Festival 239
Figini, P. 220
Figler, M.H. 108
5 V data (volume, variety, velocity,
veracity and value) 243
flâneur, concept of 223
Fluxus 223
Font, X. 80
food-based tourism 86
Fourie, J. 165
Frey, B. 184
Frost, W. 254
function volunteering: career 41;
enhancement 41; protective 41; social
41; understanding 41; values 40–41

Gagne, P. 105, 106
Gaither, C.J. 89, 99
Gartner, W.C. 244; classification of
tourist information sources 62
Gelbman, A. 173, 175
Getz, D. 3, 20, 87, 88, 92, 100, 104, 117,
118, 126, 127, 166, 167, 234, 251, 252,
254, 255
Gibson, H.J. 164
Gilmore, J.H. 26
globalization 232, 235, 236, 246
Global Muslim Travel Index
(GMTI) 73
glocalised processes 218
Goffman, K. 238, 241, 246
Google Forms 108, 110
Gouguet, J.- J. 153
Gramsci, A. 235, 237, 247
Grandi, S. 8
Grand Prix 19–20
Gray, H.P. 107
Greater Christchurch Group 132
Greater Christchurch Regeneration Act
2016 132
Green Man Festival 245
Gretzel, U. 49
Gross Domestic Product (GDP) 70–71
Gross Value Added (GVA) 234
Growth and Transformation Plan (GTP)
I & II 197
The Guardian 244

Hagstrom, M. 243
Hajubaba, H. 75
'halal tourism' *see* Islamic heritage
Hall, C.M. 127, 175
Hall, L. 172
hallmark events 170
Hamza, I.M. 74
Hanekom, D. 77
Hardcore Bikers 107
Harley-Davidson 112, 119
Harrington, S. 238
Hartmann, D. 61
Hatipoglu, B.K. 7
health tourism 6
hedonic dimensions 16
hegemony, concept of 235
Henama, U.S. 70
Henderson, J.C. 174
heritage mining 218, 219
Hernandez-Escampa, M. 8
Herrero, L.C. 185
Higgins-Desbiolles, F. 88, 101
Hobswarm, E. 19
Hodges, N. 16
Honda 112
hosting events 170
Howard, E. 15
Huang, B. 247
Hudson, S. 105
Hunter, T. 136

IBM SPSS Statistics 110
IFPI's Global Music Report 244
image-making 196
influencers 22, 52, 60, 64
Instagram 6; Ariana Grande 52;
    communities 51; followers per user
    56, *56*; image-based social network
    50; likes and comments per post
    58–59, *59*; meetups (*see* meetups,
    Instagram); participants posting
    pictures 55–58; Photo() method
    58; posts and interactivity 58–60;
    posts per user 56, *57*; as promotion
    tool 50–51; publicity of brands or
    products 50; purchasing behaviours
    50; Selena Gomez 52; smartphones,
    use of 50; UGEs 62; user faithfulness
    57–58, *58*; user profile 56–57, *57*;
    users, information about **56**; Weight()
    method 53, 60
International Congress and Convention
    Association (ICCA) 201
International Exhibitions 185

International Meetings Statistics
    Report of the Union of International
    Associations (UIA) 196
International Meeting Statistics of the
    Union of International Associations
    (UIA) 8
international motorcycle events,
    Europe 103
internet 242, 244; and social media 9,
    233; and World Wide Web 241
internet of things (IoT) 243, 244, 255
Islamic heritage 75
Islamic tourism 73; Cape Town,
    introduction of Islam 73; destinations
    74; Meccas, world pilgrimage
    destination 73–74; "Muslim
    friendliness" 74; prayer experience
    74; Robben Island, major Muslim
    heritage site 74
Isle of Man TT 106
Isle of Wight Festival 239
Israel: cultural heritage and physical
    attraction 171; destination marketing
    177; event tourism and travellers to
    171–172; Holy Land 177; religious
    tourist numbers 171; spiritual
    reflection 172; travellers to 171–172
Italy: additionality of visitors for
    Lombardy 154–155, **155**; BOP
    study **154**; estimation methods 153,
    153; event expenditure, notion of
    150; expenditure of foreigners 149;
    Expo 2015 157, *158*; hypothetical
    increase in consumption 150–153;
    Lombardy, computation of additional
    expenditure 155–156; outcome of
    analysis 150, *150*; quantification
    of additional visitors **148**;
    regional impact and inter-regional
    redistribution of wealth 153–157;
    spillover effects 157, 159; stake-
    holders' testimonies 151, **151**; visitors'
    expenditures 152, **152**, 156, 157, **157**;
    visitors spending **148**, 148–149
itinerary, concept of 222–224

Jérôme Massiani 7
job dissatisfaction 34
job experiences: at events 34–35;
    heterogeneity of 33; individual
    event 34
job satisfaction: behavioural response of
    relaxation 34; critical examination of 33;
    between employees and volunteers 35;

at events 34–35; for event staff 34; goals, achievement 34; individual's appraisal 33, 34; motivational factors 35; organisational behaviour literature 35; permanent employees 41
Johnson 189
Jones, C. 164

Kachali, H. 127
Kamish, R. 73
Kangaroo Island Surf Music Festival, case of 5
Kelly, K.G. 103, 105, 106, 108
37th Komjekejeke King Silamba Annual Commemoration 78
Komjekejeke traditional festival: community-based tourism 7, 79–81; community-initiated cultural celebration 76–77; dignitaries 77; inferences 78–79; Ndebele culture 76–77; Silamba celebrations 78–79; Wallmansthal outside Pretoria 76–78
Kozak, M. 110, 116
Kruger, M. 107, 110, 116
KwaZulu-Natal Province 75

Laconia Motorcycle Week 103
Laing, J. 254
Lamond, I.R. 235, 239
Las Vegas 20
Laven, D. 173
Lawler, E.E. 35
leisure events 169
leisure, recreation and tourism (LRT) 220
Letterist International 223
Levi-Strauss, C. 236
Lewis, J. 25
Liao, D.Y. 167
Likert-scale 110
liquid modernity 232, 237
Liutikas, D. 174
Long, P. 89
long-term branding and tourism 18
long-term sustainable development of city 219
Lovelock, B. 8
Lubbe, G. 74
Lucchetti, V.G. 80

McDowell, L. 241
McDowell, M.L. 168
MacFarlane, R.: *Psychogeography: A Beginner's Guide* 223

McGehee, N.G. 89
McGuigan 239
Mair, J. 6, 38, 127
major or mega-events 164
Malaga: classical music and orchestra concerts 192–193; construction of new museums 191; ECOC 2016 190–191; *façadism* 191; and figure of Picasso 191, 193; impact of bidding process and reactions 190–193; 'Infinite City' 191–192; Neighborhood Theatre Festival 192
Malawi wanderlust 176
Mandela, N. 74
Manyika, J. 243
Marine-Roig, E. 6
market-led policymaking 132
Marschall, S. 175
Martin-Fuentes, E. 6
Martinotti (1993) model 217
mass tourism 79
Mbeki, G. 74
meetups, Instagram 48, 51; community-based events 49; definition 49; of Instagrammers in Pyrenees (case study) 53–55, *54*; leisure and discovery events 54; qualitative analysis 55; quantitative analysis 54–55; tourism destinations and businesses 64
mega-events 170, 185; data on daily expenditures 160; impact of Expo 2015 survey (*see* Expo 2015 survey); Olympics or international exhibitions 143; substitution effects in economic analysis 144–145
Meleddu, M. 163
Memiş, S. 7
MICE (Meetings, Incentives, Conferences and Exhibitions) sector: challenges, in Addis Ababa 206–210; cities in Africa, 2016 201, **202**; destination management organizations 202; destinations in Africa, 2016 201, **202**; growth of 8; institutional structure 211; primary facilities and infrastructure 203; professional event managers 202; quality of workforce 202; sector potentials of Addis Ababa 200–206; tourism development in Ethiopia 8; tourism sector development in Ethiopia 197–200
Millennium Development Goals (MDGs) 197

Ministry of Culture and Tourism (MoCT) 198
Mitchell, J. 200
Mogensen K.A. 242
Molekwa, S. 77, 78
Monferrer, B. 185
motivational factors 35; extrinsic 35, 41; extrinsic and intrinsic 127; intrinsic 35
motorcycle events, United States: adventure motorcycling, phenomenon of 105; Africa Bike Week 107; attendees, emergence of differences 107; biker-focused initiatives 105–106; biking community 105, 119; BMW touring motorcyclists 106; clubs 106; destination choice motives 105; drive tourism typology 104; Dynamic Interaction Leisure 105; event attendance motivation 103; event attendees, travel attitudes and behaviours 103; knowledge on event tourism, framework for 104; market segmentation 106–107; motives for attending 107, **107**; motorcycle tourism 104–105; rallies 106; RTOs, creation of 105–106; tourism and tourists 104–106; tourism motivation, travellers' 103; travel characteristics and tourism motives 104; travelling to **113**, 113–114; travel motivations to **116**, 116–117; wanderlust and **114**, 114–116, **115**
Motorcycle Fests, perspective of 7
Munday, M. 164
Munich Oktoberfest 39
music festival and events industry: counterculture movements 232–233; demands of consumers 232; motivations 234; popularity of 234; potential future for 232, 245–246; process of standardised culture and dominant forces 232; role of music in society 233; social platforms 232–233
Muskat, B. 6, 40
Muslim religious festival: crisis-resistant tourists 75–76; Inferences 74–75; Ramadan in the Bo Kaap, Cape Town (*see* Ramadan)
Muslim tourism 73
Myrtle Beach Bike 103
MySpace 47
Mzansi Golden Economy (MGE) 78

Nair, B.B. 9
Nakhlistan, non-profit organisation 73

Nale, R.D. 106
National Department of Arts and Culture (DAC) 77
National Department of Tourism (NDT) 77, 78
National Sports Tourism Strategy in 2012 71
National Tourism Development Policy (NTDP) 197–198
Ndebele culture 7, 76–78
Nelson, K. 252
New Orleans 20
niche tourism 175–176; globalisation and technological advancements 175; personal memory 175; religious tourism 175
Nichols, H. 25
Nolan, D. 177
Non-Governmental Organisations (NGOs) 203
North West University (2017) 74
Novice Riders 107

Olympics 19–20, 185
"one-off" experience 3
Onishi, N. 72
online communities: destinations 65; faithfulness and sense of belonging 60; formation of 6
online photo-sharing 53
online-social phenomena 242
Oreg, S. 254
organisational ecology 129
Organization of African Unity (OAU) 203
Osland, G.E. 75
outdoor events 164
Ovacık Village Women's Seed Association 90, 92
overtourism, negative impacts of 75

Page, S.J. 88, 92, 104, 118
Palmer, R. 18, 185
Palmer/Rae Associates 187
Pan African Chamber of Commerce and Industry (PACCI) 203
Parker, B. 135; *Otago Daily Times* 135
Partal, A.S. 192
participant observation, user-generated events: basic travel motivators 60; online community, faithfulness and sense of belonging 60; participant profile 60; photography and social media, behaviour and relationship to 61

Pedersen, K.K. 242
Penfold, T. 253
Perdue, R.R. 89
Perez, M. 241
personal events 169
Peters, P. 233
photography 53; based UGEs 64; behaviour and relationship to 61; competition 63
pilgrimage management 4
Pine, B.J. 26, 252
Pinguinos (Spain) 103
Plan for Accelerated and Sustained Development to End Poverty (PASDEP) 197
planned tourism events, impacts of 170
Plymouth: America's Cup World Series Yacht Racing 2011 189; culture and tourism 189–190; election of Hull 190; impact of bidding process and reactions 189–190; Pilgrim Fathers 189, 190; 'Plymouth, Britain's Ocean City' 189, 193
Porter, L.W. 35
post-disaster Christchurch, context of 127
Price-Davies, E. 105
Pride-Festival 239
Project X (film) 47
proposed event portfolio model, DSCs: Commercial Events 29; Community Events 29; Enhancer Events 29; high impact or Attractor Events 28–29; Medium or Disruptor Events 29
psychogeography 223
pulsating nature, event organisation: event operations phase 36; post-event phase 36; pre-event phase 36

quake effects in Christchurch CBD: Christchurch Central Recovery Plan 131; Christchurch City Council (CCC) 131; damage and demolition 129–131; disruption in transport network 131; venues and conference facilities in hotels **130**

Rabinowitz, I. 106
Rajesh, R. 168
Raj, R. 164, 168
Ramadan: in Bo Kaap 7; crisis-resistant tourism destination 72–73; heritage protection for Bo Kaap 72, 81; month of fasting 73; tourism consumption (*see* Islamic tourism)

Ranger, T. 19
Raymond, E.M. 175
Reconstruction and Development Plan (RDP) 71
Regional Tourism Organizations (RTOs), creation of 105–106
religious tourism events 4, 6, 73, 163; believers and non-believers 173, 175–176, 177; decision-making process 172; faith groups and tour operators 176; heritage foods 173; management 8; pilgrimage destination 173–175; religious travellers and vacationers 174; travel marketers 172; wanderlusts and 172–175
Resident Empowerment through Tourism Scale (RETS) 89, 91
retail centres: as destinations 18–20; as fantasy palaces 16; functional shopping (utilitarian) 15; Grandview Shopping Centre in Guangzhou 14; Industry Voice: Alex Caley, Marketing And Experience Manager, Meadowhall Sheffield 17; invention of tradition 19; January 'Shopping Festival' in Dubai 14; leisurely shopping (hedonic) 15; major and mega-events 18–19; Mall of Emirates, Dubai 14; New York's new year celebrations 18; quality of service 16, 18; recreational shopping 15; shopping centres 19–20; shopping value, generation of 16; strategic outcomes 19; tourism objectives 19; unplanned and organic events 19; Visitor Attraction industry in Dubai 14–15; West Edmonton Mall in Canada 14
retail event tourism (using events to attract tourists) 20–24; advantage of using events to manage tourism, DSC 23; destination marketing 20, 22; images of events 22; infrastructure, Centres to improve 23; music festival 23–24
Return-on-Investment (ROI) 82
Richards, G. 18, 185, 234
Rio Carnival 39
Rio Olympics, Brazil 253
Ritchie, J.R. 105
Rittichainuwat, B. 127
Rodella, I. 171
Rodriguez-Gonzalez, M.A. 8
Rojo, J. 238, 239
Ron, A.S. 173, 177

rural entrepreneurship: agricultural and non-agricultural activities 89; anticipated sustainability impacts 99; cultural heritage tourism development 89; definition 88; event implementation 97–98; labor markets and distribution channels 88–89; local governmental agencies 88; needs of local community 96–97; outcomes of the event 98; partners' overlapping interests 100; planning phase 97; positive loop 98, 100; through planned event, framework for 92, *93*

Şahin, D. 7
Santana- Gallego, M. 165
Sanz, J.A. 185
Sarna, K. 240
Scheyvens, R. 7, 89, 91
Seville, E. 70, 127
Sherwood, P. 87
Shields, P.O. 108, 109–110, 117
Shinde, K.A. 176
Sigcau, S. 77
significant events 170
Şile: Earth Market 86; study setting 90–91
Şile Tourism Association 90, 92
Simeoni, U. 171
Sisulu, Walter 74
Situationist International 223
Skillen, F. 168
Slow Fish 90
Slow Food Şile-Palamut Convivium 90, 92
slow food movement: small-scale farmers and consumers 90, 95; taste education 90 (*see also* Earth Market, Şile)
Sobukwe, Robert 74
social capital 4, 5
social community 166
Socialist Revolution 197
social media 6; behaviour and relationship to 61
Social Responsibility Implementation (SRI) project 77
social value 16
Sokanyile, A. 76
Soundscape project 224
South Africa: annual Komjekejeke traditional festival 76–81; democratic transition 71; economic fortunes 70; 2010 FIFA World Cup 81; high rate

of crime 76; mining and agriculture 70–71; Muslim religious festival (*see* Ramadan); National Sports Tourism Strategy in 2012 71; Ndebele culture 76–77; "new kid on the block" 71; non-Organisation of Islamic Cooperation (OIC) destination 74; poor and unemployment 71; sustained growth 70
South African Historical Organisation (2019) 72
spatial-temporal phenomenon 127
special events 169, 185
Spencer, J.P. 70
sporting/music tourism events 20
sports 3; participation 166
sports tourism 6, 71; 2010 FIFA World Cup 6; 1995 Rugby World Cup 6
Spracklen, K. 235, 239
stakeholder agility 138
Stanciulescu, Chis, A. 172
standardisation of culture 236–237; counterculture (1960s) 245; modernization 236; notion of progressive cultural change 237; Theodor Adorno's theory of mass culture 236
Stanis, S.A.W. 175
Stevenson, J.R. 127
Strafford, D. 5
Strydom, A.J. 80
Sturgis Rally 103
substitution effects, mega-events: econometric analysis 145; *ex abrupto* 144; displacement, crowding-out 144; household expenditure surveys 145; local expenditures 144; Milan 2015 impact studies *146*; questionnaire 145
Surrealism approach 223
Survey Monkey 108; experience related to riding motorcycle 109; habits related to motorcycle events 109–110; socio-demographical data 109; travel motivations for attending motorcycle event 110, **111**; wanderlust scale 109, **109**
sustainable tourism development: assets, use of 87; complexities of planned event tourism 87–88; decision-making processes 88; democratic processes, use of 101; empowerment for 89; event outcomes 87; event tourism and 87–88; exploration of long-term sustainability impacts 101; negative

externalities for destination 87; social exchange theory 89; stakeholders and consensus-building 87; tourism events 87
Swarbrooke, J. 25
Sykes, D.M. 103, 105, 106, 108, 110

Tall Ships' Races 185
taste education 90
Teixeira, J. 188
Terry, A. 106
Timothy, D.J. 173
Tirca, A. 172
Tokarchuk, O. 163
Tomazos, K. 175
Torino 2006 Olympics, impact of 145
tourism events 163, 165; destination, critical issues in 163–171; impact of the event 163–164
Tourism Satellite Account (TSA) 210
tourism sector development, Ethiopia 197–200
Tourism Transformation Council (TTC) 198
tourism: tourist behaviour 47; tourist dissatisfaction 53; tourist gaze concept 105; tourist satisfaction 47, 53; types 71
tourist UGEs, definition 63
Toy Run 103
Travel and Tourism Competitiveness Index (TTCI) ranking 198
Trump, D. 74
TT race in the Isle of Man (United Kingdom) 103

Umar, U.M. 73–74
Unathi Sonwabile Henama 6
Underhill, P. 15
Unified Modelling Language (UML): Event class 53; Firm/DMO class 52, 53; processing of UGEs 64; to store and process an event *52*; User class 52
United Nations (UN) 203
United Nations Conference on Trade and Development 70
United Nations Economic Commission for Africa (UNECA) 196
United Nations Human Settlements Programme (UNHSP) 200
urban boosterism 126
urbanisation processes 219
urban tourism and urban regeneration 127

Urry, J. 105
user-generated content (UGC) 6; and online communities 47–48; posts, identification and analysis 52
user-generated events (UGEs) 6; analysis framework (quantitative approach) 51–53; community-based events 48–49; discussion 61–63; event tourism 49–50; Instagram, promotion tool (*see* Instagram); online image/brand dissemination 61; participant observation 60–61; qualitative analysis 55, 60–61; quantitative analysis 54–55, 55–60, 62; reproduction of tourist images 62; results 55; in tourism 51
Utizi, K. 171

Van Lennep, T. 76
Vargo, J. 127
variables, definition of 91
visitor attractions, DSC: changing face of retail 14–18; city centres 30; event portfolios as marketing strategy for DSCs 24–27; Industry Voice: Nikki Tansey, Event Manager At Intu Trafford Centre 21; proposed event portfolio model for DSC marketing strategy 28–29; retail centres as destinations 18–20; retail event tourism (using events to attract tourists) 20–24
Visser, G. 71
volunteers ("the life blood" of events) 40; experienced and inexperienced 40; function volunteering 40–41; unpaid volunteers 40, 42; Voluntary Functions Inventory (VFI) 40; younger 40

Wagner, N. 136
wanderlusts 7; data analysis 110; data collection 108–110; discussion 117–118; "impulse to travel" 108; method 110; motivation theory, evolution of 103; Motorcycle Fests, perspective of 7; motorcycle tourism and tourists (*see* motorcycle events, United States); pleasure travel, motivational factors for 108; and religious tourism events 172–175; sample description 110, 112, **112**; in tourism motivation 107–108
Watson, L. 137

Watts, C. 190
Western Cape Province 75
Westernization 232
Whitford, M. 100, 118
Whitman, Z. 70, 78
Wolf, J. 245
Wood, A. 137
Woodruffe-Burton, H. 16
Woodstock festival 239, 245
World Congress of Endometriosis 137
World Cup 19–20
World Economic Forum (WEF) 198

*World Encyclopedia of
  Entrepreneurship* 88
worldwide tourism phenomenon 219
Wright, D. 8

Yamaha 112
Yeoman, I. 36, 251, 252, 254
Yiu C. 243
Yoo, K.H. 49

Zifkos, G. 254
Žižek, S. 3

For Product Safety Concerns and Information please contact our EU
representative GPSR@taylorandfrancis.com
Taylor & Francis Verlag GmbH, Kaufingerstraße 24, 80331 München, Germany

www.ingramcontent.com/pod-product-compliance
Lightning Source LLC
Chambersburg PA
CBHW060345220326
41598CB00023B/2815

9 780367 616427